MEASURING AND SUSTAINING THE NEW ECONOMY

SOFTWARE, GROWTH, AND THE FUTURE OF THE U.S. ECONOMY

Report of a Symposium

DALE W. JORGENSON AND CHARLES W. WESSNER, EDITORS

Committee on Software, Growth, and the Future of the U.S. Economy

Committee on Measuring and Sustaining the New Economy

Board on Science, Technology, and Economic Policy

Policy and Global Affairs

NATIONAL RESEARCH COUNCIL
OF THE NATIONAL ACADEMIES

THE NATIONAL ACADEMIES PRESS
Washington, D.C.
www.nap.edu

THE NATIONAL ACADEMIES PRESS 500 Fifth Street, N.W. Washington, DC 20001

NOTICE: The project that is the subject of this report was approved by the Governing Board of the National Research Council, whose members are drawn from the councils of the National Academy of Sciences, the National Academy of Engineering, and the Institute of Medicine. The members of the committee responsible for the report were chosen for their special competences and with regard for appropriate balance.

This study was supported by: Contract/Grant No. CMRC-50SBNB9C1080 between the National Academy of Sciences and the U.S. Department of Commerce; Contract/Grant No. NASW-99037, Task Order 103, between the National Academy of Sciences and the National Aeronautics and Space Administration; Contract/Grant No. CMRC-SB134105C0038 between the National Academy of Sciences and the U.S. Department of Commerce; OFED-13416 between the National Academy of Sciences and Sandia National Laboratories; Contract/Grant No. N00014-00-G-0230, DO #23, between the National Academy of Sciences and the Department of the Navy; Contract/Grant No. NSF-EIA-0119063 between the National Academy of Sciences and the National Science Foundation; and Contract/Grant No. DOE-DE-FG02-01ER30315 between the National Academy of Sciences and the U.S. Department of Energy. Additional support was provided by Intel Corporation. Any opinions, findings, conclusions, or recommendations expressed in this publication are those of the author(s) and do not necessarily reflect the views of the organizations or agencies that provided support for the project.

International Standard Book Number 0-309-09950-1

Limited copies are available from Board on Science, Technology, and Economic Policy, National Research Council, 500 Fifth Street, N.W., W547, Washington, DC 20001; (202) 334-2200.

Additional copies of this report are available from the National Academies Press, 500 Fifth Street, N.W., Lockbox 285, Washington, DC 20055; (800) 624-6242 or (202) 334-3313 (in the Washington metropolitan area); Internet, http://www.nap.edu

Copyright 2006 by the National Academy of Sciences. All rights reserved.

Printed in the United States of America

THE NATIONAL ACADEMIES

Advisers to the Nation on Science, Engineering, and Medicine

The **National Academy of Sciences** is a private, nonprofit, self-perpetuating society of distinguished scholars engaged in scientific and engineering research, dedicated to the furtherance of science and technology and to their use for the general welfare. Upon the authority of the charter granted to it by the Congress in 1863, the Academy has a mandate that requires it to advise the federal government on scientific and technical matters. Dr. Ralph J. Cicerone is president of the National Academy of Sciences.

The **National Academy of Engineering** was established in 1964, under the charter of the National Academy of Sciences, as a parallel organization of outstanding engineers. It is autonomous in its administration and in the selection of its members, sharing with the National Academy of Sciences the responsibility for advising the federal government. The National Academy of Engineering also sponsors engineering programs aimed at meeting national needs, encourages education and research, and recognizes the superior achievements of engineers. Dr. Wm. A. Wulf is president of the National Academy of Engineering.

The **Institute of Medicine** was established in 1970 by the National Academy of Sciences to secure the services of eminent members of appropriate professions in the examination of policy matters pertaining to the health of the public. The Institute acts under the responsibility given to the National Academy of Sciences by its congressional charter to be an adviser to the federal government and, upon its own initiative, to identify issues of medical care, research, and education. Dr. Harvey V. Fineberg is president of the Institute of Medicine.

The **National Research Council** was organized by the National Academy of Sciences in 1916 to associate the broad community of science and technology with the Academy's purposes of furthering knowledge and advising the federal government. Functioning in accordance with general policies determined by the Academy, the Council has become the principal operating agency of both the National Academy of Sciences and the National Academy of Engineering in providing services to the government, the public, and the scientific and engineering communities. The Council is administered jointly by both Academies and the Institute of Medicine. Dr. Ralph J. Cicerone and Dr. Wm. A. Wulf are chair and vice chair, respectively, of the National Research Council.

www.national-academies.org

Committee on Software, Growth, and the Future of the U.S. Economy*

Dale W. Jorgenson, *Chair*
Samuel W. Morris University Professor
Harvard University

Kenneth Flamm
Dean Rusk Chair in International Affairs
LBJ School of Public Affairs
University of Texas at Austin

Jack Harding
Chairman, President, and CEO
eSilicon Corporation

Monica S. Lam
Professor of Computer Science
Stanford University

William Raduchel

Anthony E. Scott
Chief Information Technology Officer
General Motors

William J. Spencer
Chairman Emeritus, *retired*
International SEMATECH

Hal R. Varian
Class of 1944 Professor
School of Information Management and Systems
University of California at Berkeley

*As of February 2004.

Committee on Measuring and Sustaining the New Economy*

Dale Jorgenson, *Chair*
Samuel W. Morris University Professor
Harvard University

William J. Spencer, *Vice Chair*
Chairman Emeritus, *retired*
International SEMATECH

M. Kathy Behrens
Managing Director of Medical Technology
Robertson Stephens Investment Management

Kenneth Flamm
Dean Rusk Chair in International Affairs
LBJ School of Public Affairs
University of Texas at Austin

Bronwyn Hall
Professor of Economics
University of California at Berkeley

James Heckman
Henry Schultz Distinguished Service
Professor of Economics
University of Chicago

Ralph Landau
Consulting Professor of Economics
Stanford University

Richard Levin
President
Yale University

David T. Morgenthaler
Founding Partner
Morgenthaler Ventures

Mark B. Myers
Visiting Executive Professor of Management
The Wharton School
University of Pennsylvania

Roger Noll
Morris M. Doyle
Centennial Professor of Economics
Stanford University

Edward E. Penhoet
Chief Program Officer
Science and Higher Education
Gordon and Betty Moore Foundation

William Raduchel

Alan Wm. Wolff
Managing Partner
Dewey Ballantine

*As of February 2004.

Project Staff*

Charles W. Wessner
Study Director

Sujai J. Shivakumar
Program Officer

Ken Jacobson
Consultant

McAlister T. Clabaugh
Program Associate

David E. Dierksheide
Program Associate

---

*As of February 2004.

For the National Research Council (NRC), this project was overseen by the Board on Science, Technology and Economic Policy (STEP), a standing board of the NRC established by the National Academies of Sciences and Engineering and the Institute of Medicine in 1991. The mandate of the STEP Board is to integrate understanding of scientific, technological, and economic elements in the formulation of national policies to promote the economic well-being of the United States. A distinctive characteristic of STEP's approach is its frequent interactions with public- and private-sector decision makers. STEP bridges the disciplines of business management, engineering, economics, and the social sciences to bring diverse expertise to bear on pressing public policy questions. The members of the STEP Board* and the NRC staff are listed below:

Dale Jorgenson, *Chair*
Samuel W. Morris University Professor
Harvard University

M. Kathy Behrens
Managing Director of Medical Technology
Robertson Stephens Investment Management

Bronwyn Hall
Professor of Economics
University of California at Berkeley

James Heckman
Henry Schultz Distinguished Service
Professor of Economics
University of Chicago

Ralph Landau
Consulting Professor of Economics
Stanford University

Richard Levin
President
Yale University

William J. Spencer, *Vice Chair*
Chairman Emeritus, *retired*
International SEMATECH

David T. Morgenthaler
Founding Partner
Morgenthaler

Mark B. Myers
Visiting Executive Professor of Management
The Wharton School
University of Pennsylvania

Roger Noll
Morris M. Doyle
Centennial Professor of Economics
Stanford University

Edward E. Penhoet
Chief Program Officer
Science and Higher Education
Gordon and Betty Moore Foundation

William Raduchel

Alan Wm. Wolff
Managing Partner
Dewey Ballantine

*As of February 2004.

STEP Staff*

Stephen A. Merrill
Executive Director

Charles W. Wessner
Program Director

Russell Moy
Senior Program Officer

Sujai J. Shivakumar
Program Officer

Craig M. Schultz
Research Associate

David E. Dierksheide
Program Associate

McAlister T. Clabaugh
Program Associate

---

*As of February 2004.

Contents

Preface

Significant and sustained increases in semiconductor productivity, predicted by Moore's Law, has ushered a revolution in communications, computing, and information management.[1] This technological revolution is linked to a distinct rise in the mid 1990s of the long-term growth trajectory of the United States.[2] Indeed, U.S. productivity growth has accelerated in recent years, despite a series of negative economic shocks. Analysis by Dale Jorgenson, Mun Ho, and Kevin Stiroh of the sources of this growth over the 1996 to 2003 period suggests that the production and use of information technology account for a large share of the gains. The authors go further to project that during the next decade, private-sector

[1]This is especially so for the computer hardware sector and perhaps for the Internet as well, although there is insufficient empirical evidence on the degree to which the Internet may be responsible. For a discussion of the impact of the Internet on economic growth see, "A Thinker's Guide," *The Economist*, March 30, 2000. For a broad study of investment in technology-capital and its use in various sectors, see McKinsey Global Institute, *U.S. Productivity Growth 1995-2000: Understanding the Contribution of Information Technology Relative to Other Factors,* Washington, D.C.: McKinsey & Co., October 2001.

[2]See Dale W. Jorgenson and Kevin J. Stiroh, "Raising the Speed Limit: U.S. Economic Growth in the Information Age," in National Research Council, *Measuring and Sustaining the New Economy,* Dale W. Jorgenson and Charles W. Wessner, eds., Washington, D.C.: National Academy Press, 2002.

productivity growth will continue at a rate of 2.6 percent per year.[3] The New Economy is, thus, not a fad, but a long-term productivity shift of major significance.[4]

The idea of a "New Economy" brings together the technological innovations, structural changes, and public policy challenges associated with measuring and sustaining this remarkable economic phenomenon.

- Technological innovation—more accurately, the rapid rate of technological innovation in information technology (including computers, software, and telecommunications) and the rapid growth of the Internet—are now widely seen as underpinning the productivity gains that characterize the New Economy.[5] These productivity gains derive from greater efficiencies in the production of computers from expanded use of information technologies.[6] Many therefore believe that the productivity growth of the New Economy draws from the technological innovations found in information technology industries.[7]
- Structural changes arise from a reconfiguration of knowledge networks and business patterns made possible by innovations in information technology. Phenomena, such as business-to-business e-commerce and Internet retailing, are altering how firms and individuals interact, enabling greater efficiency in pur-

---

[3]Dale W. Jorgenson, Mun S. Ho, and Kevin J. Stiroh, "Will the U.S. Productivity Resurgence Continue?" *FRBNY Current Issues in Economics and Finance,* 10(1), 2004.

[4]The introduction of advanced productivity-enhancing technologies obviously does not eliminate the business cycle. See Organisation for Economic Co-operation and Development, *Is There a New Economy? A First Report on the OECD Growth Project,* Paris: Organisation for Economic Co-operation and Development, June 2000, p. 17. See also, M.N. Baily and R.Z. Lawrence, "Do We Have an E-conomy?" NBER Working Paper 8243, April 23, 2001, at *<http://www.nber.org/papers/w8243>*.

[5]Broader academic and policy recognition of the New Economy can be seen, for example, from the "Roundtable on the New Economy and Growth in the United States" at the 2003 annual meetings of the American Economic Association, held in Washington, D.C. Roundtable participants included Martin Baily, Martin Feldstein, Robert J. Gordon, Dale Jorgenson, Joseph Stiglitz, and Lawrence Summers. Even those who were initially skeptical about the New Economy phenomenon now find that the facts support the belief that faster productivity growth has proved more durable and has spread to other areas of the economy—e.g., retail, banking. See *The Economist*, "The new 'new economy,' " September 11, 2003.

[6]See, for example, Stephen Oliner and Daniel Sichel, "The Resurgence of Growth in the late 1990's: Is Information Technology the Story?" *Journal of Economic Perspectives,* 14(4) Fall 2000. Oliner and Sichel estimate that improvements in the computer industry's own productive processes account for about a quarter of the overall productivity increase. They also note that the use of information technology by all sorts of companies accounts for nearly half the rise in productivity.

[7]See Alan Greenspan's remarks before the White House Conference on the New Economy, Washington D.C., April 5, 2000, *<http://www.federalreserve.gov/BOARDDOCS/SPEECHES/2000/20000405.HTM>*. For a historical perspective, see the Proceedings of this volume. Ken Flamm compares the economic impact of semiconductors today with the impact of railroads in the nineteenth century.

chases, production processes, and inventory management.[8] Offshore outsourcing of service production is another manifestation of structural changes made possible by new information and communications technologies. These structural changes are still emerging as the use and applications of the Internet continue to evolve.

• Public policy plays a major role at several levels. This includes the government's role in fostering rules of interaction within the Internet[9] and its discretion in setting and enforcing the rules by which technology firms, among others, compete.[10] More familiarly, public policy concerns particular fiscal and regulatory choices that can affect the rate and focus of investments in sectors such as telecommunications. The government also plays a critical role within the innovation system.[11] It provides national research capacities,[12] incentives to promote education and training in critical disciplines, and funds most of the nation's basic research.[13] The government also plays a major role in stimulating innovation, most broadly through the patent system.[14] Government procurement and awards

---

[8]See, for example, Brookes Martin and Zaki Wahhaj, "The Shocking Economic Impact of B2B" *Global Economic Paper*, 37, Goldman Sachs, February 3, 2000.

[9]Dr. Vint Cerf notes that the ability of individuals to interact in potentially useful ways within the infrastructure of the still expanding Internet rests on its basic rule architecture: "The reason it can function is that all the networks use the same set of protocols. An important point is these networks are run by different administrations, which must collaborate both technically and economically on a global scale." See comments by Dr. Cerf in National Research Council, *Measuring and Sustaining the New Economy, op. cit.* Also in the same volume, see the presentation by Dr. Shane Greenstein on the evolution of the Internet from academic and government-related applications to the commercial world.

[10]The relevance of competition policy to the New Economy is manifested by the intensity of interest in the antitrust case, *United States versus Microsoft*, and associated policy issues.

[11]See Richard Nelson, ed., *National Innovation Systems*, New York: Oxford University Press, 1993.

[12]The STEP Board has recently completed a major review of the role and operation of government-industry partnerships for the development of new technologies. See National Research Council, *Government-Industry Partnerships for the Development of New Technologies: Summary Report*, Charles W. Wessner, ed., Washington, D.C.: National Academies Press, 2002.

[13]National Research Council, *Trends in Federal Support of Research in Graduate Education,* Stephen A. Merrill, ed., Washington, D.C.: National Academy Press, 2001.

[14]In addition to government-funded research, intellectual property protection plays an essential role in the continued development of the biotechnology industry. See Wesley M. Cohen and John Walsh, "Public Research, Patents and Implications for Industrial R&D in the Drug, Biotechnology, Semiconductor and Computer Industries" in National Research Council, *Capitalizing on New Needs and New Opportunities: Government-Industry Partnerships in Biotechnology and Information Technologies*, Washington, D.C.: National Academy Press, 2001. There is a similar situation in information technology with respect to the combination of generally non-appropriable government-originated innovation and appropriable industry intellectual property creation. The economic rationale for government investment is based on the non-appropriablity of many significant information technology innovations, including the most widely used idiomatic data structures and algorithms, as well as design and architectural patterns. Also, the IT industry relies on a number of technical and process commonalities or standards such as the suite of Internet protocols, programming languages, core design patterns, and architectural styles.

also encourage the development of new technologies to fulfill national missions in defense, health, and the environment.[15]

Collectively, these public policies play a central role in the development of the New Economy. Sustaining this New Economy will require public policy to remain relevant to the rapid technological and structural changes that characterize it. This is particularly important because of the "unbounded" nature of information technology that underpins the New Economy. Information technology and software production are not commodities that the United States can potentially afford to give up overseas suppliers but, as William Raduchel noted in his workshop presentation, a part of the economy's production function. This characteristic means that a loss of U.S. leadership in information technology and software will damage, in an ongoing way, the nation's future ability to compete in diverse industries, not least the information technology industry. Collateral consequences of a failure to develop adequate policies to sustain national leadership in information technology is likely to extend to a wide variety of sectors from financial services and health care to telecom and automobiles, with critical implications for our nation's security and the well-being of our citizens.

THE CONTEXT OF THIS REPORT

Since 1991 the National Research Council's Board on Science, Technology, and Economic Policy (STEP) has undertaken a program of activities to improve policy-makers' understanding of the interconnections between science, technology, and economic policy and their importance to the American economy and its international competitive position. The Board's interest in the New Economy and its underpinnings derive directly from its mandate.

This mandate has previously been reflected in STEP's widely cited volume, *U.S. Industry in 2000,* which assesses the determinants of competitive performance in a wide range of manufacturing and service industries, including those

---

[15]For example, government support played a critical role in the early development of computers. See Kenneth Flamm, *Creating the Computer,* Washington, D.C.: The Brookings Institution, 1988. For an overview of government industry collaboration, see the introduction to the recent report on the Advanced Technology Program, National Research Council, *The Advanced Technology Program: Assessing Outcomes,* Charles W. Wessner, ed., Washington, D.C.: National Academy Press, 2001. The historical and technical case for government-funded research in IT is well documented in reports by the Computer Science and Telecommunications Board (CSTB) of the National Research Council. In particular, see National Research Council, *Innovation in Information Technology,* Washington, D.C.: National Academies Press, 2003. This volume provides an update of the "tire tracks" diagram first published in CSTB's 1995 Brooks-Sutherland report, which depicts the critical role that government-funded university research has played in the development of the multibillion-dollar IT industry.

relating to information technology.[16] The Board also undertook a major study, chaired by Gordon Moore of Intel, on how government-industry partnerships can support growth-enhancing technologies.[17] Reflecting a growing recognition of the importance of the surge in productivity since 1995, the Board launched a multifaceted assessment, exploring the sources of growth, measurement challenges, and the policy framework required to sustain the New Economy. The first exploratory volume was published in 2002.[18] Subsequent workshops and ensuing reports in this series include *Productivity and Cyclicality in the Semiconductor Industry* and *Deconstructing the Computer.* The present report, *Software, Growth, and the Future of the U.S. Economy,* examines the role of software and its importance to U.S. productivity growth; how software is made and why it is unique; the measurement of software in national and business accounts; the implications of the movement of the U.S. software industry offshore; and related policy issues.

SYMPOSIUM AND DISCUSSIONS

Believing that increased productivity in the semiconductor, computer component, and software industries plays a key role in sustaining the New Economy, the Committee on Measuring and Sustaining the New Economy, under the auspices of the STEP Board, convened a symposium February 20, 2004, at the National Academy of Sciences, Washington, D.C. The symposium on *Software, Growth, and the Future of the U.S. Economy* drew together expertise from leading academics, national accountants, and innovators in the information technology sector (Appendix B lists these individuals).

The "Proceedings" chapter of this volume contains summaries of their workshop presentations and discussions. Also included in this volume is a paper by William Raduchel on "The Economics of Software," which was presented at the symposium. Given the quality and the number of presentations, summarizing the workshop proceedings has been a challenge. We have made every effort to capture the main points made during the presentations and the ensuing discussions. We apologize for any inadvertent errors or omissions in our summary of the proceedings. The lessons from this symposium and others in this series will contribute to the Committee's final consensus report on *Measuring and Sustaining the New Economy.*

---

[16] National Research Council, *U.S. Industry in 2000: Studies in Competitive Performance,* David C. Mowery, ed., Washington, D.C.: National Academy Press, 1999.

[17] For a summary of this multi-volume study, See National Research, *Government-Industry Partnerships for the Development of New Technologies: Summary Report, op. cit.*

[18] National Research Council, *Measuring and Sustaining the New Economy: Report of a Workshop, op. cit.*

ACKNOWLEDGMENTS

There is considerable interest in the policy community in developing a better understanding of the technological drivers and appropriate regulatory framework for the New Economy, as well as in a better grasp of its operation. This interest is reflected in the support on the part of agencies that have played a role in the creation and development of the New Economy. We are grateful for the participation and the contributions of the National Aeronautical and Space Administration, the Department of Energy, the National Institute of Standards and Technology, the National Science Foundation, and Sandia National Laboratories.

We are indebted to Ken Jacobson for his preparation of the meeting summary. Several members of the STEP staff also deserve recognition for their contributions to the preparation of this report. We wish to thank Sujai Shivakumar for his contributions to the introduction to the report. We are also indebted to McAlister Clabaugh and David Dierksheide for their role in preparing the conference and getting this report ready for publication.

NRC REVIEW

This report has been reviewed in draft form by individuals chosen for their diverse perspectives and technical expertise, in accordance with procedures approved by the National Academies' Report Review Committee. The purpose of this independent review is to provide candid and critical comments that will assist the institution in making its published report as sound as possible and to ensure that the report meets institutional standards for quality and objectivity. The review comments and draft manuscript remain confidential to protect the integrity of the process.

We wish to thank the following individuals for their review of this report: Bruce Grimm, Bureau of Economic Analysis; Shane Greenstein, Northwestern University; David Messerschmitt, University of California, Berkeley; William Scherlis, Carnegie Mellon University; and Andrew Viterbi, Viterbi Group LLC.

Although the reviewers listed above have provided many constructive comments and suggestions, they were not asked to endorse the content of the report, nor did they see the final draft before its release. The review of this report was overseen by Robert White, Carnegie Mellon University. Appointed by the National Academies, he was responsible for making certain that an independent examination of this report was carried out in accordance with institutional procedures and that all review comments were carefully considered. Responsibility for the final content of this report rests entirely with the authors and the institution.

STRUCTURE

This report has three parts: an Introduction; a summary of the proceedings of the February 20, 2004, symposium; and a research paper by Dr. William Raduchel. Finally, a bibliography provides additional references.

This report represents an important step in a major research effort by the Board on Science, Technology, and Economic Policy to advance our understanding of the factors shaping the New Economy, the metrics necessary to understand it better, and the policies best suited to sustaining the greater productivity and prosperity that it promises.

Dale W. Jorgenson　　　　Charles W. Wessner

I

INTRODUCTION

Software and the New Economy

The New Economy refers to a fundamental transformation in the United States economy as businesses and individuals capitalize on new technologies, new opportunities, and national investments in computing, information, and communications technologies. Use of this term reflects a growing conviction that widespread use of these technologies makes possible a sustained increase in the productivity and growth of the U.S. economy.[1]

Software is an encapsulation of knowledge in an executable form that allows for its repeated and automatic applications to new inputs.[2] It is the means by which we interact with the hardware underpinning information and communications technologies. Software is increasingly an integral and essential part of most goods and services—whether it is a handheld device, a consumer appliance, or a retailer. The United States economy, today, is highly dependent on software with

[1]In the context of this analysis, the New Economy does not refer to the boom economy of the late 1990s. The term is used in this context to describe the acceleration in U.S. productivity growth that emerged in the mid-1990s, in part as a result of the acceleration of Moore's Law and the resulting expansion in the application of lower cost, higher performance information technologies. See Dale W. Jorgenson, Kevin J. Stiroh, Robert J. Gordon, and Daniel E. Sichel, "Raising the Speed Limit: U.S. Economic Growth in the Information Age," *Brookings Papers on Economic Activity*, (1):125-235, 2000.

[2]See the presentation by Monica Lam, summarized in the Proceedings section of this volume.

businesses, public utilities, and consumers among those integrated within complex software systems.

- Almost every aspect of a modern corporation's operations is embodied in software. According to Anthony Scott of General Motors, a company's software embodies a whole corporation's knowledge into business processes and methods—"virtually everything we do at General Motors has been reduced in some fashion or another to software."
- Much of our public infrastructure relies on the effective operation of software, with this dependency also leading to significant vulnerabilities. As William Raduchel observed, it seems that the failure of one line of code, buried in an energy management system from General Electric, was the initial source leading to the electrical blackout of August 2003 that paralyzed much of the northeastern and midwestern United States.[3]
- Software is also redefining the consumer's world. Microprocessors embedded in today's automobiles require software to run, permitting major improvements in their performance, safety, and fuel economy. And new devices such as the iPod are revolutionizing how we play and manage music, as personal computing continues to extend from the desktop into our daily activities.

As software becomes more deeply embedded in most goods and services, creating reliable and robust software is becoming an even more important challenge.

Despite the pervasive use of software, and partly because of its relative immaturity, understanding the economics of software presents an extraordinary challenge. Many of the challenges relate to measurement, econometrics, and industry structure. Here, the rapidly evolving concepts and functions of software as well as its high complexity and context-dependent value makes measuring software difficult. This frustrates our understanding of the economics of software—both generally and from the standpoint of action and impact—and impedes both policy making and the potential for recognizing technical progress in the field.

While the one-day workshop gathered a variety of perspectives on software, growth, measurement, and the future of the New Economy, it of course could not (and did not) cover every dimension of this complex topic. For example, workshop participants did not discuss the potential future opportunities in leveraging software in various application domains. This major topic considers the potential for major future opportunities for software to revolutionize key sectors of the U.S. economy, including the health care industry. The focus of the meeting was on developing a better understanding of the economics of software.

Indeed, as Dale Jorgenson pointed out in introducing the National Academies conference on Software and the New Economy, "we don't have a very

[3] "Software Failure Cited in August Blackout Investigation," *Computerworld*, November 20, 2003.

clear understanding collectively of the economics of software." Accordingly, a key goal of this conference was to expand our understanding of the economic nature of software, review how it is being measured, and consider public policies to improve measurement of this key component of the nation's economy, and measures to ensure that the United States retains its lead in the design and implementation of software.

Introducing the Economics of Software, Dr. Raduchel noted that software pervades our economy and society.[4] As Dr. Raduchel further pointed out, software is not merely an essential market commodity but, in fact, embodies the economy's production function itself, providing a platform for innovation in all sectors of the economy. This means that sustaining leadership in information technology (IT) and software is necessary if the United States is to compete internationally in a wide range of leading industries—from financial services, health care and automobiles to the information technology industry itself.

MOORE'S LAW AND THE NEW ECONOMY

The National Academies' conference on software in the New Economy follows two others that explored the role of semiconductors and computer components in sustaining the New Economy. The first National Academies conference in the series considered the contributions of semiconductors to the economy and the challenges associated with maintaining the industry on the trajectory anticipated by Moore's Law. Moore's Law anticipates the doubling of the number of transistors on a chip every 18 to 24 months. As Figure 1 reveals, Moore's Law has set the pace for growth in the capacity of memory chips and logic chips from 1970 to 2002.[5]

An economic corollary of Moore's Law is the fall in the relative prices of semiconductors. Data from the Bureau of Economic Analysis (BEA), depicted in Figure 2, shows that semiconductor prices have been declining by about 50 percent a year for logic chips and about 40 percent a year for memory chips between 1977 and 2000. This is unprecedented for a major industrial input. According to Dale Jorgenson, this increase in chip capacity and the concurrent fall in price—the "faster-cheaper" effect—created powerful incentives for firms to substitute information technology for other forms of capital, leading to the productivity increases that are the hallmark of the New Economy.[6]

The second National Academies conference on the New Economy, "Deconstructing the Computer," examined how the impact of Moore's Law and

[4]See William Raduchel, "The Economics of Software," in this volume.

[5]For a review of Moore's Law on its fortieth anniversary, see *The Economist*, "Moore's Law at 40" March 26, 2005.

[6]Dale W. Jorgenson "Information Technology and the U.S. Economy," *The American Economic Review*, 91(1):1-32, 2001.

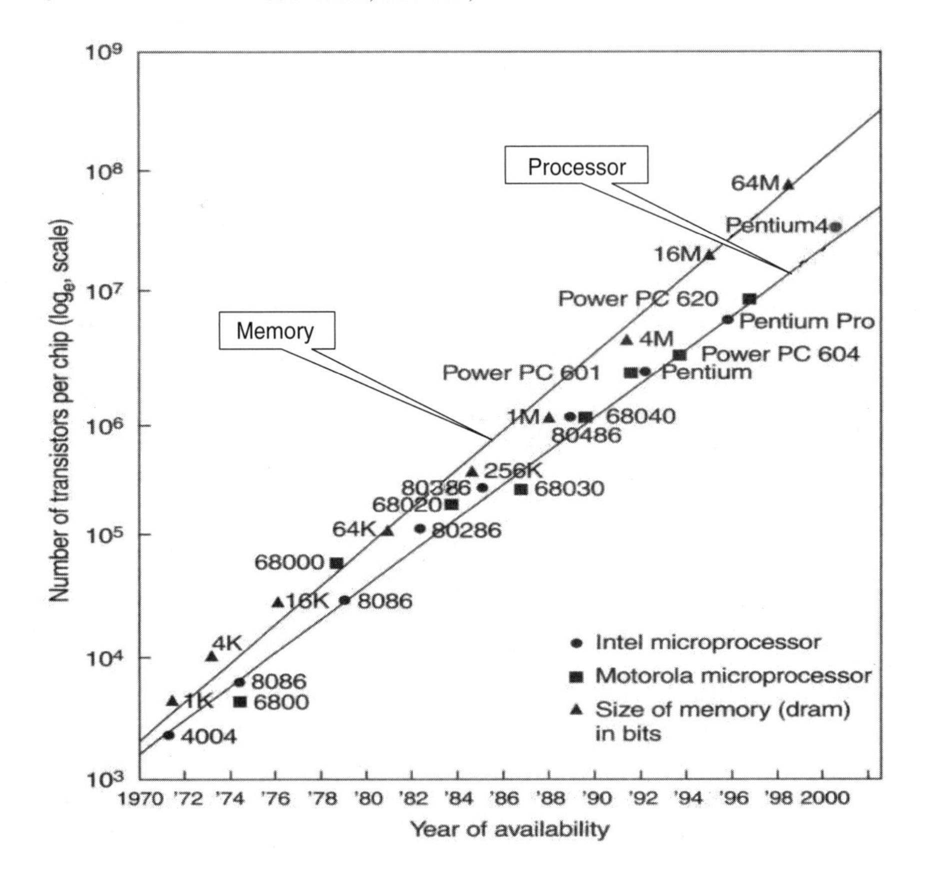

FIGURE 1 Transistor density on microprocessors and memory chips.

its price corollary extended from microprocessors and memory chips to high technology hardware such as computers and communications equipment. While highlighting the challenges of measuring the fast-evolving component industries, that conference also brought to light the impact of computers on economic growth based on BEA price indexes for computers (See Figure 3). These figures reveal that computer prices have declined at about 15 percent per year between 1977 and the new millennium, helping to diffuse modern information technology across a broad spectrum of users and applications.

The New Economy is alive and well today. Recent figures indicate that, since the end of the last recession in 2001, information-technology-enhanced productivity growth has been running about two-tenths of a percentage point higher than

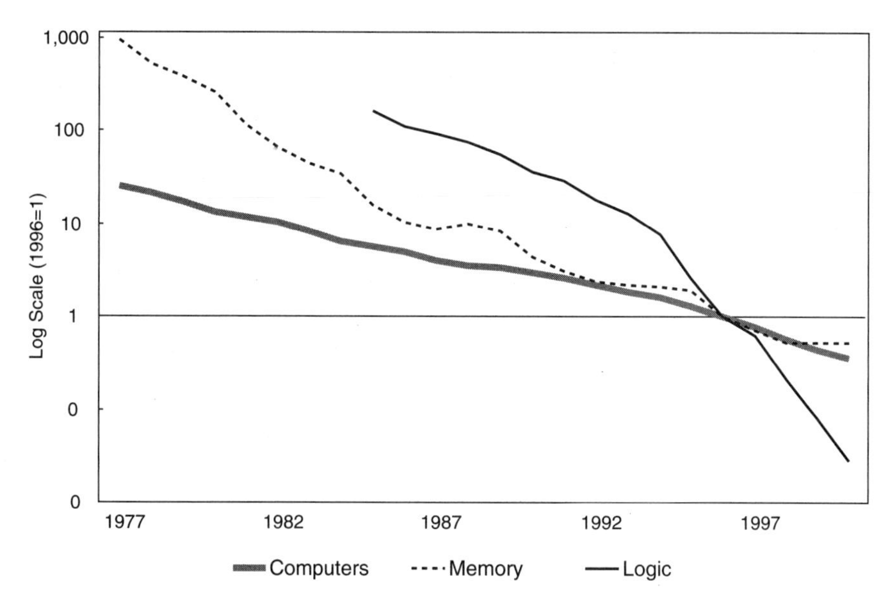

FIGURE 2 Relative prices of computers and semiconductors, 1977-2000.
NOTE: All price indexes are divided by the output price index.

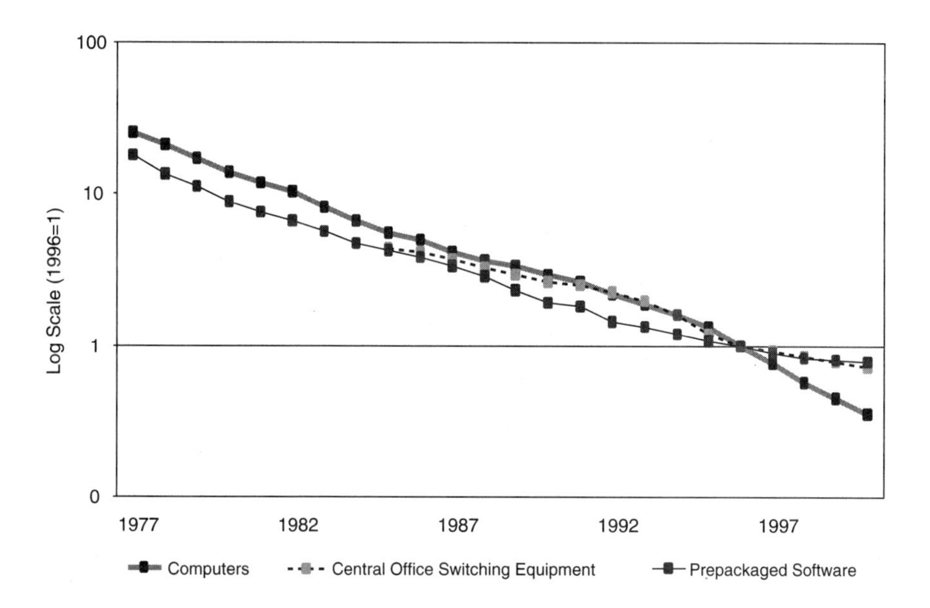

FIGURE 3 Relative prices of computers, communications, and software, 1977-2000.
NOTE: All price indexes are divided by the output price index.

in any recovery of the post-World War II period.[7] The current challenge rests in developing evidence-based policies that will enable us to continue to enjoy the fruits of higher productivity in the future. It is with this aim that the Board on Science, Technology, and Economic Policy of the National Academies has undertaken a series of conferences to address the need to measure the parameters of the New Economy as an input to better policy making and to highlight the policy challenges and opportunities that the New Economy offers.

This volume reports on the third National Academies conference on the New Economy and Software.[8] While software is generally believed to have become more sophisticated and more affordable over the past three decades, data to back these claims remains incomplete. BEA data show that the price of prepackaged software has declined at rates comparable to those of computer hardware and communications equipment (See Figure 3). Yet *prepackaged software* makes up only about 25 to 30 percent of the software market. There remain large gaps in our knowledge about *custom software* (such as those produced by SAP or Oracle for database management, cost accounting, and other business functions) and *own-account software* (which refers to special purpose software such as for airlines reservations systems and digital telephone switches). There also exists some uncertainty in classifying software, with distinctions made among prepackaged, custom, and own-account software often having to do more with market relationships and organizational roles rather than purely technical attributes.[9]

In all, as Dale Jorgenson points out, there is a large gap in our understanding of the sources of growth in the New Economy.[10] Consequently, a major purpose of the third National Academies conference was to draw attention to the need to

---

[7]Dale Jorgenson, Mun Ho, and Kevin Stiroh, "Will the U.S. Productivity Resurgence Continue?" *FRBNY Current Issues in Economics and Finance*, 10(13), 2004.

[8]The National Academies series on the New Economy concluded with a conference on the role of telecommunications equipment (which relies heavily on software.) The "faster-better-cheaper" phenomenon—associated with Moore's Law—is also evident in the telecom sector, which has seen enormous increases in capacity of telecommunications equipment combined with rapidly declining quality-adjusted prices. See National Research Council, *The Telecommunications Challenge: Changing Technologies and Evolving Policies*, Dale W. Jorgenson and Charles W. Wessner, eds., Washington, D.C.: National Academies Press, forthcoming.

[9]One pattern, most typical for Lotus Notes for example, is a three-way relationship whereby a consumer organization acquires a vendor system and simultaneously hires a consultant organization to configure and manage that system, due to its complexity and the "capital cost" of the learning curve. Sometimes that consultant organization is a service-focused division of the vendor, and sometimes it is a third party with appropriate licenses and training certifications from the vendor.

[10]One source of confusion is the vagueness of the definition of the software industry. For example, some believe that the financial sector spends more developing software than the software vendor sector. This suggests that IT-driven growth is caused by IT adoption generally and not just by the products provided specifically by the IT industry.

address this gap in our ability to understand and measure the trends and contribution of software to the operation of the American economy.

THE NATURE OF SOFTWARE

To develop a better economic understanding of software, we first need to understand the nature of software itself. Software, comprising of millions of lines of code, operates within a *stack*. The stack begins with the *kernel*, which is a small piece of code that talks to and manages the hardware. The kernel is usually included in the *operating system*, which provides the basic services and to which all programs are written. Above this operating system is *middleware,* which "hides" both the operating system and the window manager. For the case of desktop computers, for example, the operating system runs other small programs called *services* as well as specific *applications* such as Microsoft Word and PowerPoint. Thus, when a desktop computer functions, the entire stack is in operation. This means that the value of any part of a software stack depends on how it operates within the rest of the stack.[11]

The stack itself is highly complex. According to Monica Lam of Stanford University, software may be the most intricate thing that humans have learned to build. Moreover, it is not static. Software grows more complex as more and more lines of code accrue to the stack, making software engineering much more difficult than other fields of engineering. With hundreds of millions of lines of code making up the applications that run a big company, for example, and with those applications resting on middleware and operating systems that, in turn, comprise tens of millions of lines of code, the average corporate IT system today is far more complicated than the Space Shuttle, says William Raduchel.

The way software is built also adds to its complexity and cost. As Anthony Scott of GM noted, the process by which corporations build software is "somewhat analogous to the Winchester Mystery House," where accretions to the stack over time create a complex maze that is difficult to fix or change.[12] This complexity means that a failure manifest in one piece of software, when added to the stack, may not indicate that something is wrong with that piece of software *per se*, but quite possibly can cause the failure of some other piece of the stack that is being tested for the first time in conjunction with the new addition.

---

[11]Other IT areas have their own idiomatic "stack" architectures. For example, there are more CPUs in industrial control systems, than on desktops, and these embedded systems do not have "window managers." A similar point can be made for mainframe systems, distributed systems, and other non-desktop computing configurations.

[12]The Winchester Mystery House, in San Jose, California, was built by the gun manufacturer heiress who believed that she would die if she stopped construction on her house. Ad hoc construction, starting in 1886 and continuing over nearly four decades with no master architectural plan, created an unwieldy mansion with a warren of corridors and staircases that often lead nowhere.

Box A: The Software Challenge

According to Dr. Raduchel a major challenge in creating a piece of code lies in figuring out how to make it "error free, robust against change, and capable of scaling reliably to incredibly high volumes while integrating seamlessly and reliably to many other software systems in real time."

Other challenges involved in a software engineering process include cost, schedule, capability/features, quality (dependability, reliability, security), performance, scalability/flexibility, and many others. These attributes often involve trade offs against one another, which means that priorities must be set. In the case of commercial software, for example, market release deadlines may be a primary driver, while for aerospace and embedded health devices, software quality may be the overriding priority.

Writing and Testing Code

Dr. Lam described the software development process as one comprising various iterative stages. (She delineated these stages for analytical clarity, although they are often executed simultaneously in modern commercial software production processes.) After getting an idea of the requirements, software engineers develop the needed architecture and algorithms. Once this high-level design is established, focus shifts to coding and testing the software. She noted that those who can write software at the kernel level are a very limited group, perhaps numbering only the hundreds worldwide. This reflects a larger qualitative difference among software developers, where the very best software developers are orders of magnitude—up to 20 to 100 times—better than the average software developer.[13] This means that a disproportionate amount of the field's creative work is done by a surprisingly small number of people.

As a rule of thumb, producing software calls for a ratio of one designer to 10 coders to 100 testers, according to Dr. Raduchel.[14] Configuring, testing, and tuning the software account for 95 to 99 percent of the cost of all software in operation. These non-linear complementarities in the production of software mean that simply adding workers to one part of the production process is not likely to make a software project finish faster.[15] Further, since a majority of time in devel-

[13]This is widely accepted folklore in the software industry, but one that is difficult to substantiate because it is very difficult to measure software engineering productivity.

[14]This represents Dr. Raduchel's estimate. Estimates vary in the software industry.

[15]See Fred Brooks, *The Mythical Man Month,* New York: Addison-Wesley, 1975.

oping a software program deals with handling exceptions and in fixing bugs, it is often hard to estimate software development time.[16]

The Economics of Open-source Software

Software is often developed in terms of a stack, and basic elements of this stack can be developed on a proprietary basis or on an open or shared basis.[17] According to Hal Varian, *open-source* is, in general, software whose source code is freely available for use or modification by users and developers (and even hackers.) By definition, open-source software is different from proprietary software whose makers do not make the source code available. In the real world, which is always more complex, there are a wide range of access models for open-source software, and many proprietary software makers provide "open"-source access to their product but with proprietary markings.[18] While open-source software is a public good, there are many motivations for writing open-source software, he added, including (at the edge) scratching a creative itch and demonstrating skill to one's peers. Indeed, while ideology and altruism provide some of the motivation, many firms, including IBM, make major investments in Linux and other open-source projects for solid market reasons.

While the popular idea of a distributed model of open-source development is one where spontaneous contributions from around the world are merged into a functioning product, most successful distributed open-source developments take place within pre-established or highly precedented architectures. It should thus not come as a surprise that open source has proven to be a significant and successful way of creating robust software. Linux provides a major instance where both a powerful standard and a working reference for implementation have appeared at the same time. Major companies, including Amazon.com and Google, have chosen Linux as the kernel for their software systems. Based on this kernel, these companies customize software applications to meet their particular business needs.

Dr. Varian added that a major challenge in developing open-source software is the threat of "forking" or "splintering." Different branches of software can arise from modifications made by diverse developers in different parts of the

---

[16]There are econometric models for cost estimation in specific domains. See for example, Barry Boehm, Chris Abts, A. Brown, Sunita Chulani, Bradford Clark, and Ellis Horowitz, *Software Cost Estimation with Cocomo II,* Pearson Education, 2005.

[17]Not all software is developed in terms of a stack. Indeed, modern e-commerce frameworks have very different structures.

[18]Sun, for example, provides source access to its Java Development Kit and now Solaris, but the code is not "open source." Microsoft provides Windows source access to some foreign governments and enterprise customers.

world and a central challenge for any open-source software project is to maintain an interchangeable and interoperable standard for use and distribution. Code forks increase adoption risk for users due to the potential for subsequent contrary tips, and thus can diminish overall market size until adoption uncertainties are reduced. Modularity and a common designers' etiquette—adherence to which is motivated by the developer's self-interest in avoiding the negative consequences of forking—can help overcome some of these coordination problems.

Should the building of source code by a public community be encouraged? Because source code is an incomplete representation of the information associated with a software system, some argue that giving it away is good strategy. Interestingly, as Dr. Raduchel noted, open-source software has the potential to be much more reliable and secure than proprietary software,[19] adding that the open-source software movement could also serve as an alternative and counterweight to monopoly proprietorship of code, such as by Microsoft, with the resulting competition possibly spurring better code writing.[20] The experience in contributing to open-source also provides important training in the art of software development, he added, helping to foster a highly specialized software labor force. Indeed, the software design is a creative design process at every level—from low-level code to overall system architecture. It is rarely routine because such routine activities in software are inevitably automated.

Dr. Varian suggested that software is most valuable when it can be combined, recombined, and built upon to produce a secure base upon which additional applications can in turn be built.[21] The policy challenge, he noted, lies in ensuring the existence of incentives that sufficiently motivate individuals to develop robust basic software components through open-source coordination, while ensuring that, once they are built, they will be widely available at low cost so that future development is stimulated.

---

[19]While measurement techniques vary, successful open-source and mainstream commercial quality techniques have been found to be essentially similar and yield similar results. Indeed, open-source quality was considered lower for a while, but that was measurement error due primarily to choices made by early open-source project leaders regarding the timing and frequency of releases, not absolute achievable levels of quality. These release times, for major projects, have since been adjusted by open-source leaders to conform to more mainstream practices. See, for example, T. J. Halloran and William Scherlis, "High Quality and Open Source Software Practices," Position Paper, 24th International Conference on Software Engineering, 2002.

[20]In addition to market factors creating incentives to produce better software, software quality can also depend on the extent to which available tools, techniques, and organizational systems permit software developers to diagnose and correct errors.

[21]The great success of reuse (building the secure base) is evident in the widespread adoption of e-commerce frameworks (.NET, J2EE), GUI frameworks (Swing, MFC), and other major libraries and services. Additionally, nearly all modern information systems are based on vendor-provided core databases, operating systems, and file systems that have the benefit of billions of dollars of "capitalization" and large user communities.

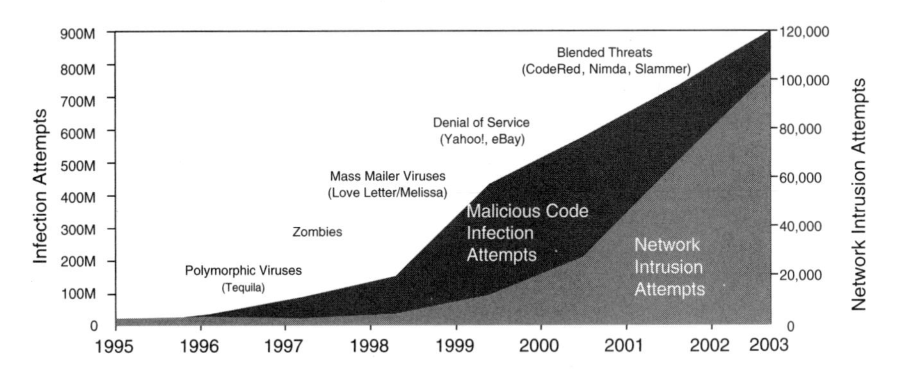

FIGURE 4 Attacks against code are growing.
SOURCE: Analysis by Symantec Security Response using data from Symantec, IDC & ICSA.

Software Vulnerability and Security—A Trillion Dollar Problem?

As software has become more complex, with a striking rise in the lines of code over the past decade, attacks against that code—in the form of both network intrusions and infection attempts—have also grown substantially[22] (See Figure 4).

The perniciousness of the attacks is also on the rise. The Mydoom attack of January 28, 2004, for example, did more than infect individuals' computers, producing acute but short-lived inconvenience. It also reset the machine's settings leaving ports and doorways open to future attacks.

The economic impact of such attacks is increasingly significant. According to Kenneth Walker of Sonic Wall, Mydoom and its variants infected up to half a million computers. The direct impact of the worm includes lost productivity owing to workers' inability to access their machines, estimated at between $500 and $1,000 per machine, and the cost of technician time to fix the damage. According to one estimate cited by Mr. Walker, Mydoom's global impact by February 1, 2004 was alone $38.5 billion. He added that the *E-Commerce Times* had estimated the global impact of worms and viruses in 2003 to be over one trillion dollars.

To protect against such attacks, Mr. Walker advocated a system of layered security, analogous to the software stack, which would operate from the network gateway, to the servers, to the applications that run those servers. Such protection, which is not free, is another indirect cost of software vulnerability that is typically borne by the consumer.

[22]For a review of the problem of software vulnerability to cyber-attacks, see Howard F. Lipson, "Tracking and Tracing Cyber-Attacks: Technical Challenges and Global Policy Issues," CERT Coordination Center, CMU/SEI2002-SR-009, November 2002.

Enhancing Software Reliability

Acknowledging that software will never be error free and fully secure from attack or failure, Dr. Lam suggested that the real question is not whether these vulnerabilities can be eliminated, raising instead the issue of the role of incentives facing software makers to develop software that is more reliable.

One factor affecting software reliability is the nature of market demand for software. Some consumers—those in the market for mass client software, for example—may look to snap up the latest product or upgrade and feature add-ons, placing less emphasis on reliability. By contrast, more reliable products can typically be found in markets where consumers are more discerning, such as in the market for servers.

Software reliability is also affected by the relative ease or difficulty in creating and using metrics to gauge quality. Maintaining valid metrics can be highly challenging given the rapidly evolving and technically complex nature of software. In practice, software engineers often rely on measurements of highly indirect surrogates for quality (relating to such variables as teams, people, organizations, processes) as well as crude size measures (such as lines of code and raw defect counts).

Other factors that can affect software reliability include the current state of liability law and the unexpected and rapid development of a computer hacker culture, which has significantly raised the complexity of software and the threshold of software reliability. While for these and other reasons it is not realistic to expect a 100 percent correct program, Dr. Lam noted that the costs and consequences of this unreliability are often passed on to the consumer.

THE CHALLENGE OF MEASURING SOFTWARE

The unique nature of software poses challenges for national accountants who are interested in data that track software costs and aggregate investment in software and its impact on the economy. This is important because over the past 5 years investment in software has been about 1.8 times as large as private fixed investment in computers peripheral equipment, and was about one-fifth of all private fixed investment in equipment and software.[23] Getting a good measure of this asset, however, is difficult because of the unique characteristics of software development and marketing: Software is complex; the market for software is different from that of other goods; software can be easily duplicated, often at low cost; and the service life of software is often hard to anticipate.[24] Even so, repre-

[23]Bureau of Economic Analysis, National Income and Product Income Table 5.3.5 on Private Fixed Investment by Type.

[24]For example, the lifespan for some Defense Department software can be two decades or more in length.

sentatives from the BEA and the Organisation for Economic Co-operation and Development (OECD) described important progress that is being made in developing new accounting rules and surveys to determine where investment in software is going, how much software is being produced in the United States, how much is being imported, and how much the country is exporting.

Financial Reporting and Software Data

Much of our data about software comes from information that companies report to the Securities and Exchange Commission (SEC). These companies follow the accounting standards developed by the Financial Accounting Standards Board (FASB).[25] According to Shelly Luisi of the SEC, the FASB developed these accounting standards with the investor, and not a national accountant, in mind. The Board's mission, after all, is to provide the investor with unbiased information for use in making investment and credit decisions, rather than a global view of the industry's role in the economy.

Outlining the evolution of the FASB's standards on software, Ms. Luisi recounted that the FASB's 1974 Statement of Financial Accounting Standards (FAS-2) provided the first standard for capitalizing software on corporate balance sheets. FAS-2 has since been developed though further interpretations and clarifications. FASB Interpretation No. 6, for instance, recognized the development of software as R&D and drew a line between software for sale and software for operations. In 1985, FAS-86 introduced the concept of technological feasibility, seeking to identify that point where the software project under development qualifies as an asset, providing guidance on determining when the cost of software development can be capitalized. In 1998, FASB promulgated "Statement of Position 98-1" that set a different threshold for capitalization for the cost of software for internal use—one that allows it to begin in the design phase, once the preliminary project state is completed and a company commits to the project. As a result of these accounting standards, she noted, software is included as property, plant, and equipment in most financial statements rather than as an intangible asset.

Given these accounting standards, how do software companies actually recognize and report their revenue? Speaking from the perspective of software company, Greg Beams of Ernst & Young noted that while sales of prepackaged software is generally reported at the time of sale, more complex software systems require recurring maintenance to fix bugs and to install upgrades, causing revenue reporting to become more complicated. In light of these multiple deliverables, software companies come up against rules requiring that they allocate value to each of those deliverables and then recognize revenue in accordance with the

[25]The Financial Accounting Standards Board is a private organization that establishes standards of financial accounting and reporting governing the preparation of financial reports. It is officially recognized as authoritative by the Securities and Exchange Commission.

requirements for those deliverables. How this is put into practice results in a wide difference in when and how much revenue is recognized by the software company—making it, in turn, difficult to understand the revenue numbers that a particular software firm is reporting.

Echoing Ms. Luisi's caveat, Mr. Beams noted that information published in software vendors' financial statements is useful mainly to the shareholder. He acknowledged that detail is often lacking in these reports, and that distinguishing one software company's reporting from another and aggregating such information so that it tells a meaningful story can be extremely challenging.

Gauging Private Fixed Software Investment

Although the computer entered into commercial use some four decades earlier, the BEA has recognized software as a capital investment (rather than as an intermediate expense) only since 1999. Nevertheless, Dr. Jorgenson noted that there has been much progress since then, with improved data to be available soon.

Before reporting on this progress, David Wasshausen of the BEA identified three types of software used in national accounts: *Prepackaged* (or shrink-wrapped) software is packaged, mass-produced software. It is available off-the-shelf, though increasingly replaced by online sales and downloads over the Internet. In 2003, the BEA placed business purchases of prepackaged software at around $50 billion. *Custom* software refers to large software systems that perform business functions such as database management, human resource management, and cost accounting.[26] In 2003, the BEA estimates business purchases of custom software at almost $60 billion. Finally, *own-account* software refers to software systems built for a unique purpose, generally a large project such as an airlines reservations system or a credit card billing system. In 2003, the BEA estimated business purchases of own-account software at about $75 billion.

Dr. Wasshausen added that the BEA uses the "commodity flow" technique to measure prepackaged and custom software. Beginning with total receipts, the BEA adds imports and subtracts exports, which leaves the total available domestic supply. From that figure, the BEA subtracts household and government purchases to come up with an estimate for aggregate business investment in software.[27] By contrast, BEA calculates own-account software as the sum of production costs,

[26]The line between prepackaged and custom software is not always distinct. National accountants have to determine, for example, whether Oracle 10i, which is sold in a product-like fashion with a license, is to be categorized as custom or prepackaged software.

[27]The BEA compares demand-based estimates for software available from the U.S. Census Bureau's Capital Expenditure Survey with the supply-side approach of the commodity flow technique. The Census Bureau is working to expand its survey to include own-account software and other information not previously captured, according to David Wasshausen. See remarks by Dr. Wasshausen in the Proceedings section of this volume.

including compensation for programmers and systems analysts and such intermediate inputs as overhead, electricity, rent, and office space.[28]

The BEA is also striving to improve the quality of its estimates, noted Dr. Wasshausen. While the BEA currently bases its estimates for prepackaged and custom software on trended earnings data from corporate reports to the SEC, Dr. Wasshausen hoped that the BEA would soon benefit from Census Bureau data that capture receipts from both prepackaged and custom software companies through quarterly surveys. Among recent BEA improvements, Dr. Wasshausen cited an *expansion of the definitions* of prepackaged and custom software imports and exports, and *better estimates* of how much of the total prepackaged and custom software purchased in the United States was for intermediate consumption. The BEA, he said, was also looking forward to an improved Capital Expenditure Survey by the Census Bureau.

Dirk Pilat of the OECD noted that methods for estimating software investment have been inconsistent across the countries of the OECD. One problem contributing to the variation in measures of software investment is that the computer services industry represents a heterogeneous range of activities, including not only software production, but also such things as consulting services. National accountants have had differing methodological approaches (for example, on criteria determining what should be capitalized) leading to differences between survey data on software investment and official measures of software investments as they show up in national accounts.

Attempting to mend this disarray, Dr. Pilat noted that the OECD Eurostat Task Force has published its recommendations on the use of the commodity flow model and on how to treat own-account software in different countries.[29] He noted that steps were underway in OECD countries to harmonize statistical practices and that the OECD would monitor the implementation of the Task Force recommendations. This effort would then make international comparisons possible, resulting in an improvement in our ability to ascertain what was moving where—the "missing link" in addressing the issue of offshore software production.

Despite the comprehensive improvements in the measurement of software undertaken since 1999, Dr. Wasshausen noted that accurate software measurement continued to pose severe challenges for national accountants simply because software is such a rapidly changing field. He noted in this regard, the rise of demand computing, open-source code development, and overseas outsourcing,

---

[28]The BEA's estimates for own-account are derived from employment and mean wage data from the Bureau of Labor Statistic's Occupational Employment Wage Survey and a ratio of operating expenses to annual payroll from the Census Bureau's Business Expenditures Survey.

[29]Organisation for Economic Co-operation and Development, *Statistics Working Paper 2003/1: Report of the OECD Task Force on Software Measurement in the National Accounts*, Paris: Organisation for Economic Co-operation and Development, 2003.

Box B: The Economist's Challenge: Software as a Production Function

Software is "the medium through which information technology expresses itself," says William Raduchel. Most economic models miscast software as a machine, with this perception dating to the period 40 years ago when software was a minor portion of the total cost of a computer system. The economist's challenge, according to Dr. Raduchel, is that software is not a factor of production like capital and labor, but actually embodies the production function, for which no good measurement system exists.

which create new concepts, categories, and measurement challenges.[30] Characterizing attempts made so far to deal with the issue measuring the New Economy as "piecemeal"—"we are trying to get the best price index for software, the best price index for hardware, the best price index for LAN equipment routers, switches, and hubs"—he suggested that a single comprehensive measure might better capture the value of hardware, software, and communications equipment in the national accounts. Indeed, information technology may be thought of as a "package," combining hardware, software, and business-service applications.

Tracking Software Prices Changes

Another challenge in the economics of software is tracking price changes. Incorporating computer science and computer engineering into the economics of software, Alan White and Ernst Berndt presented their work on estimating price changes for prepackaged software, based on their assessment of Microsoft Corporation data.[31] Dr. White noted several important challenges facing those seeking to construct measures of price and price change. One challenge lies in ascertaining which price to measure, since software products may be sold as full

[30]For example, how is a distinction to be made between service provisioning (sending data to a service outsource) and the creation and use of a local organizational asset (sending data to a service application internally developed or acquired)? The user experience may be identical (e.g., web-based access) and the geographic positioning of the server (e.g., at a secure remote site, with geography unknown to the individual user) may also be identical. In other words, the technology and user experience both look almost the same, but the contractual terms of provisioning are very different.

[31]Jaison R. Abel, Ernst R. Berndt, Alan G. White, "Price Indexes for Microsoft's Personal Computer Software Products," NBER Working Paper 9966, Cambridge, MA: National Bureau for Economic Research, 2003. The research was originally sponsored by Microsoft Corporation, though the authors are responsible for its analysis.

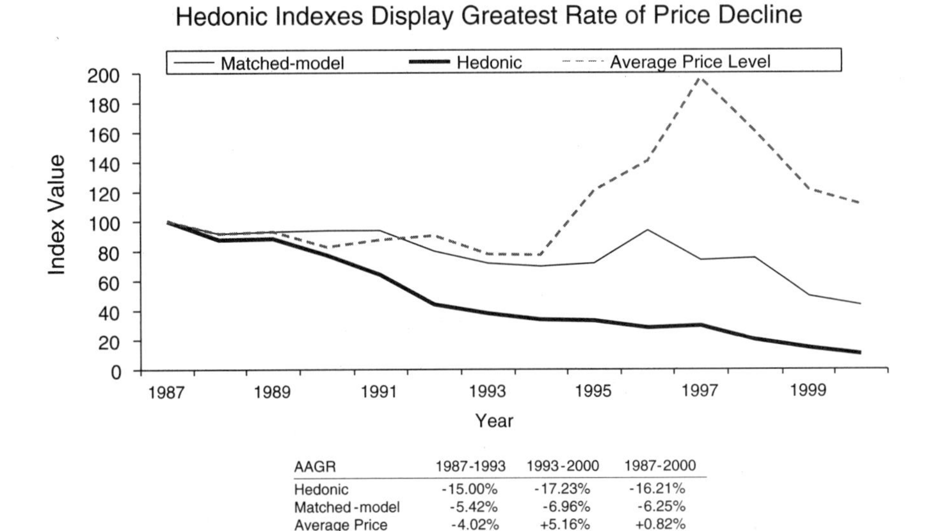

AAGR	1987-1993	1993-2000	1987-2000
Hedonic	-15.00%	-17.23%	-16.21%
Matched-model	-5.42%	-6.96%	-6.25%
Average Price	-4.02%	+5.16%	+0.82%

FIGURE 5 Quality-adjusted prices for operating systems have fallen, 1987-2000.
SOURCE: Jaison R. Abel, Ernst R. Berndt, Cory W. Monroe, and Alan G. White, "Hedonic Price Indexes for Operating Systems and Productivity Suite PC Software," draft working paper, 2004.

versions or as upgrades, stand-alones, or suites. An investigator has also to determine what the unit of output is, how many licenses there are, and when price is actually being measured. Another key issue concerns how the quality of software has changed over time and how that should be incorporated into price measures.

Surveying the types of quality changes that might come into consideration, Dr. Berndt gave the example of improved graphical interface and "plug-n-play," as well as increased connectivity between different components of a software suite. In their study, Drs. White and Berndt compared the *average* price level (computing the price per operating system as a simple average) with *quality-adjusted* price levels using hedonic and matched-model econometric techniques. They found that while the average price, which does not correct for quality changes, showed a growth rate of about 1 percent a year, the matched model showed a price decline of around 6 percent a year and the hedonic calculation showed a much larger price decline of around 16 percent.

These quality-adjusted price declines for software operating systems, shown in Figure 5, support the general thesis that improved and cheaper information technologies contributed to greater information technology adoption leading to productivity improvements characteristic of the New Economy.[32]

[32]Jorgenson et al., 2000, *op cit.*

THE CHALLENGES OF THE SOFTWARE LABOR MARKET

A Perspective from Google

Wayne Rosing of Google said that about 40 percent of the company's thousand plus employees were software engineers, which contributed to a company culture of "designing things." He noted that Google is "working on some of the hardest problems in computer science…and that someday, anyone will be able to ask a question of Google and we'll give a definitive answer with a great deal of background to back up that answer." To meet this goal, Dr. Rosing noted that Google needed to pull together "the best minds on the planet and get them working on these problems."

Google, he noted is highly selective, hiring around 300 new workers in 2003 out of an initial pool of 35,000 resumes sent in from all over the world. While he attributed this high response to Google's reputation as a good place to work, Google in turn looked for applicants with high "raw intelligence," strong computer algorithm skills and engineering skills, and a high degree of self-motivation and self-management needed to fit in with Google's corporate culture.

Google's outstanding problem, Dr. Rosing lamented, was that "there aren't enough good people" available. Too few qualified computer science graduates were coming out of American schools, he said. While the United States remained one of the world's top areas for computer-science education and produced very good graduates, there are not enough people graduating at the Masters' or Doctoral level to satisfy the needs of the U.S. economy, especially for innovative firms such as Google. At the same time, Google's continuing leadership requires having capable employees from around the world, drawing on advances in technology and providing language specific skills to service various national markets. As a result, Google hires on a global basis.

A contributing factor to Google's need to hire engineers outside the country, he noted, is the impact of U.S. visa restrictions. Noting that the H1-B quota for 2004 was capped at 65,000, down from approximately 225,000 in previous years, he said that Google was not able to hire people who were educated in the United States, but who could not stay on and work for lack of a visa. Dr. Rosing said that such policies limited the growth of companies like Google within the nation's borders—something that did not seem to make policy sense.

The Offshore Outsourcing Phenomenon

Complexity and efficiency are the drivers of offshore outsourcing, according to Jack Harding of eSilicon, a relatively new firm that produces custom-made microchips. Mr. Harding noted that as the manufacturing technology grows more complex, a firm is forced to stay ahead of the efficiency curve through large recapitalization investments or "step aside and let somebody else do that part of

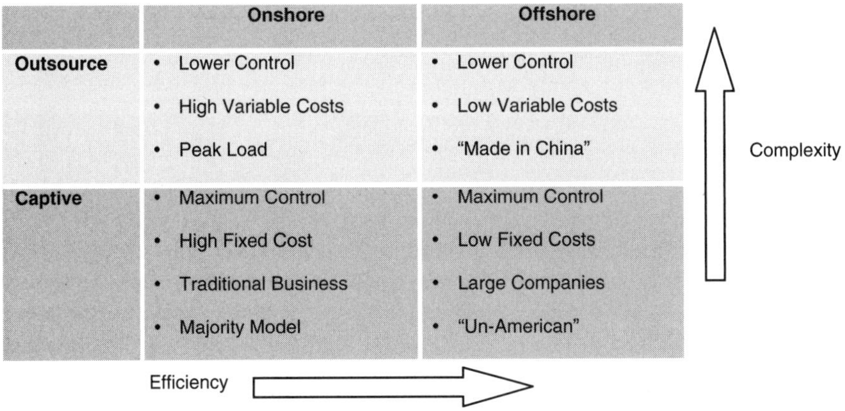

	Onshore	Offshore
Outsource	• Lower Control • High Variable Costs • Peak Load	• Lower Control • Low Variable Costs • "Made in China"
Captive	• Maximum Control • High Fixed Cost • Traditional Business • Majority Model	• Maximum Control • Low Fixed Costs • Large Companies • "Un-American"

Complexity

Efficiency

FIGURE 6 The Offshore Outsourcing Matrix.

the work." This decision to move from captive production to outsourced production, he said, can then lead to offshore-outsourcing—or "offshoring"—when a company locates a cheaper supplier in another country of same or better quality.

Displaying an outsourcing-offshoring matrix (Figure 6) Mr. Harding noted that it was actually the "Captive-Offshoring" quadrant, where American firms like Google or Oracle open production facilities overseas, that is the locus of a lot of the current "political pushback" about being "un-American" to take jobs abroad.[33] Activity that could be placed in the "Outsource-Offshore" box, meanwhile, was marked by a trade-off where diminished corporate control had to be weighed against very low variable costs with adequate technical expertise.[34]

Saving money by outsourcing production offshore not only provides a compelling business motive, it has rapidly become "best practice" for new companies. Though there might be exceptions to the rule, Mr. Harding noted that a software company seeking venture money in Silicon Valley that did not have a plan to base a development team in India would very likely be disqualified. It would not be seen as competitive if its intention was to hire workers at $125,000 a year in Silicon Valley when comparable workers were available for $25,000 a year in Bangalore. (See Figure 7, cited by William Bonvillian, for a comparison of annual salaries for software programmers.) Heeding this logic, almost every

[33]For example, see the summary of remarks by Mr. James Socas, captured in the Proceedings section of this volume.

[34]Although Mr. Harding distinguished between "captive offshoring" and "offshore outsourcing," most speakers used the term "offshore outsourcing" or "offshoring" less precisely to refer to both phenomena in general. We follow the general usage in the text here.

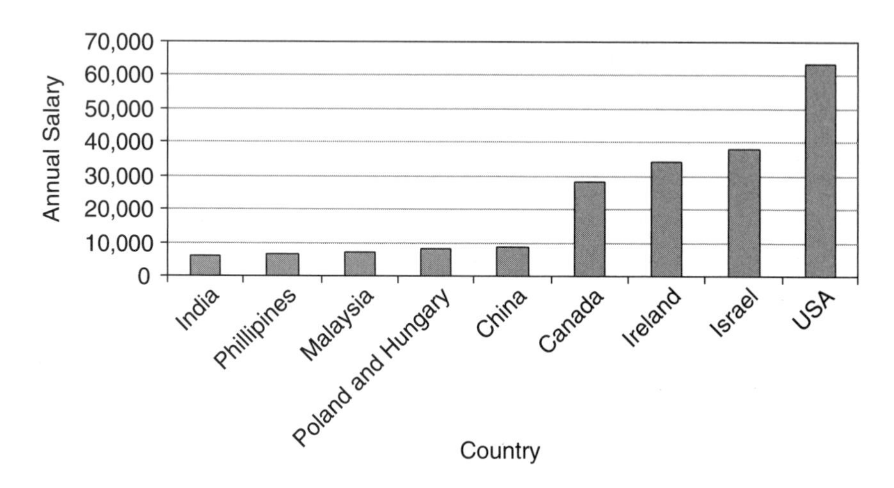

FIGURE 7 Averages of base annual salary for a programmer in various countries.
SOURCE: *Computerworld*, April 28, 2003.

software firm has moved or is in the process of moving its development work to locations like India, observed Mr. Harding. The strength of this business logic, he said, made it imperative that policy-makers in the United States understand that offshoring is irreversible and learn how to constructively deal with it.

How big is the offshoring phenomenon? Despite much discussion, some of it heated, the scope of the phenomenon is poorly documented. As Ronil Hira of the Rochester Institute of Technology pointed out, the lack of data means that no one could say with precision, how much work had actually moved offshore. This is clearly a major problem from a policy perspective.[35] He noted, however, that the effects of these shifts were palpable from the viewpoint of computer hardware engineers and electrical and electronics engineers, whose ranks had faced record levels of unemployment in 2003.

SUSTAINING THE NEW ECONOMY: THE IMPACT OF OFFSHORING

What is the impact of the offshoring phenomenon on the United States and what policy conclusions can we draw from this assessment? Conference participants offered differing, often impassioned views on this question. Presenters did

[35]Nevertheless, Dr. Hira implied that job loss and downward wage pressures in the United States information technology sector were related to the employment of hardware and software engineers abroad. In his presentation, Dr. Hira noted that he is the chair of the Career and Workforce Committee of IEEE-USA, a professional society that represents 235,000 U.S. engineers.

not agree with one another, even on such seemingly simple issues, such as whether the H1-B quota is too high or too low, whether the level for H1-B visas is causing U.S. companies to go abroad for talent or not, whether there is a shortage of talent within U.S. borders or not, or whether there is a short-term over-supply (or not) of programmers in the present labor market.

Conference participants, including Ronil Hira and William Bonvillian, highlighted two schools of thought on the impact of offshore outsourcing—both of which share the disadvantage of inadequate data support. Whereas some who take a macroeconomic perspective believe that offshoring will yield lower product and service costs and create new markets abroad fueled by improved local living standards, others, including some leading industrialists who understand the micro implications, have taken the unusual step of arguing that offshoring can erode the United States' technological competitive advantage and have urged constructive policy countermeasures.

Among those with a more macro outlook, noted Dr. Hira, is Catherine Mann of the Institute for International Economics, who has argued that "just as for IT hardware, globally integrated production of IT software and services will reduce these prices and make tailoring of business-specific packages affordable, which will promote further diffusion of IT use and transformation throughout the U.S.

Box C: Two Contrasting Views on Offshore Outsourcing

"Outsourcing is just a new way of doing international trade. More things are tradable than were in the past and that's a good thing. . . . I think that outsourcing is a growing phenomenon, but it's something that we should realize is probably a plus for the economy in the long run."

N. Gregory Mankiw[a]

"When you look at the software industry, the market share trend of the U.S.-based companies is heading down and the market share of the leading foreign companies is heading up. This x-curve mirrors the development and evolution of so many industries that it would be a miracle if it didn't happen in the same way in the IT service industry. That miracle may not be there."

Andy Grove

[a]Dr. Mankiw made this remark in February 2004, while Chairman of the President's Council of Economic Advisors. Dr. Mankiw drew a chorus of criticism from Congress and quickly backpedaled, although other leading economists supported him. See "Election Campaign Hit More Sour Notes," *The Washington Post*, p. F-02, February 22, 2004.

economy."[36] Cheaper information technologies will lead to wider diffusion of information technologies, she notes, sustaining productivity enhancement and economic growth.[37] Dr. Mann acknowledges that some jobs will go abroad as production of software and services moves offshore, but believes that broader diffusion of information technologies throughout the economy will lead to an even greater demand for workers with information technology skills.[38]

Observing that Dr. Mann had based her optimism in part on the unrevised Bureau of Labor Statistics (BLS) occupation projection data, Dr. Hira called for reinterpreting this study in light of the more recent data. He also stated his disagreement with Dr. Mann's contention that lower IT services costs provided the only explanation for either rising demand for IT products or the high demand for IT labor witnessed in the 1990s. He cited as contributing factors the technological paradigm shifts represented by such major developments as the growth of the Internet as well as Object-Oriented Programming and the move from mainframe to client-server architecture.

Dr. Hira also cited a recent study by McKinsey and Company that finds, with similar optimism, that offshoring can be a "win-win" proposition for the United States and countries like India that are major loci of offshore outsourcing for software and services production.[39] Dr. Hira noted, however, that the McKinsey estimates relied on optimistic estimates that have not held up to recent job market realities. McKinsey found that India gains a net benefit of at least 33 cents from every dollar the United States sends offshore, while America achieves a net benefit of at least $1.13 for every dollar spent, although the model apparently assumes that India buys the related products from the United States.

These more sanguine economic scenarios must be balanced against the lessons of modern growth theorists, warned William Bonvillian in his conference presentation. Alluding to Clayton Christiansen's observation of how successful companies tend to swim upstream, pursuing higher-end, higher-margin customers

---

[36]Catherine Mann, "Globalization of IT Services and White Collar Jobs: The Next Wave of Productivity Growth," *International Economics Policy Briefs,* PB03-11, December 2003.

[37]Lael Brainard and Robert Litan have further underlined the benefits to the United States economy, in this regard, noting that lower inflation and higher productivity, made possible through offshore outsourcing, can allow the Federal Reserve to run a more accommodative monetary policy, "meaning that overall and over time the [U.S.] economy will grow faster, creating the conditions for higher overall employment." See Lael Brainard and Robert E. Litan, " 'Offshoring' Service Jobs: Bane or Boon and What to Do?" *Brookings Institution Policy Brief* 132, April 2004.

[38]Challenging the mainstream economics consensus about the benefits of offshore outsourcing, Prof. Samuelson asserts that the assumption that the laws of economics dictate that the U.S. economy will benefit from all forms of international trade is a "popular polemical untruth." See Paul Samuelson, "Where Ricardo and Mill Rebut and Confirm Arguments of Mainstream Economists Supporting Globalization" *Journal of Economic Perspectives* 18(3), 2004.

[39]McKinsey Global Institute, "Offshoring: Is it a win-win game?" San Francisco: McKinsey Global Institute, 2003.

Box D: Software, Public Policy, and National Competitiveness

Information technology and software production are not commodities that the United States can potentially afford to give up overseas suppliers but are, as William Raduchel noted in his workshop presentation, a part of the economy's production function (See Box B). This characteristic means that a loss of U.S. leadership in information technology and software will damage, in an ongoing way, the nation's future ability to compete in diverse industries, not least the information technology industry. Collateral consequences of a failure to develop adequate policies to sustain national leadership in information technology is likely to extend to a wide variety of sectors from financial services and health care to telecom and automobiles, with critical implications for our nation's security and the well-being of Americans.

with better technology and better products, Mr. Bonvillian noted that nations can follow a similar path up the value chain.[40] Low-end entry and capability, made possible by outsourcing these functions abroad, he noted, can fuel the desire and capacity of other nations to move to higher-end markets.

Acknowledging that a lack of data makes it impossible to track activity of many companies engaging in offshore outsourcing with any precision, Mr. Bonvillian noted that a major shift was underway. The types of jobs subject to offshoring are increasingly moving from low-end services—such as call centers, help desks, data entry, accounting, telemarketing, and processing work on insurance claims, credit cards, and home loans—towards higher technology services such as software and microchip design, business consulting, engineering, architecture, statistical analysis, radiology, and health care where the United States currently enjoys a comparative advantage.

Another concern associated with the current trend in offshore outsourcing is the future of innovation and manufacturing in the United States. Citing Michael Porter and reflecting on Intel Chairman Andy Grove's concerns, Mr. Bonvillian noted that business leaders look for locations that gather industry-specific resources together in one "cluster." [41] Since there is a tremendous skill set involved in

[40]Clayton Christiansen, *The Innovator's Dilemma: When New Technologies Cause Great Firms to Fail*, Cambridge: Harvard Business School Press, 1997.

[41]Michael Porter, "Building the Microeconomic Foundations of Prosperity: Findings from the Business Competitiveness Index," *The Global Competitiveness Report 2003-2004*, X Sala-i-Martin, ed., New York: Oxford University Press, 2004.

advanced technology, argued Mr. Bonvillian, losing a parts of that manufacturing to a foreign country will help develop technology clusters abroad while hampering their ability to thrive in the United States. These effects are already observable in semiconductor manufacturing, he added, where research and development is moving abroad to be close to the locus of manufacturing.[42] This trend in hardware, now followed by software, will erode the United States' comparative advantage in high technology innovation and manufacture, he concluded.

The impact of these migrations is likely to be amplified: Yielding market leadership in software capability can lead to a loss of U.S. software advantage, which means that foreign nations have the opportunity to leverage their relative strength in software into leadership in sectors such as financial services, health care, and telecom, with potentially adverse impacts on national security and economic growth.

Citing John Zysman, Mr. Bonvillian pointed out that "manufacturing matters," even in the Information Age. According to Dr. Zysman, advanced mechanisms for production and the accompanying jobs are a strategic asset, and their location makes the difference as to whether or not a country is an attractive place to innovate, invest, and manufacture.[43] For the United States, the economic and strategic risks associated with offshoring, noted Mr. Bonvillian, include a loss of in-house expertise and future talent, dependency on other countries on key technologies, and increased vulnerability to political and financial instabilities abroad.

With data scarce and concern "enormous" at the time of this conference, Mr. Bonvillian reminded the group that political concerns can easily outstrip economic analysis. He added that a multitude of bills introduced in Congress seemed to reflect a move towards a protectionist outlook.[44] After taking the initial step of collecting data, he noted that lawmakers would be obliged to address widespread public concerns on this issue. Near-term responses, he noted, include programs to retrain workers, provide job-loss insurance, make available additional venture financing for innovative startups, and undertake a more aggressive trade policy. Longer term responses, he added, must focus on improving the nation's innovative capacity by investing in science and engineering education and improving the broadband infrastructure.

---

[42]National Research Council, *Securing the Future, Regional and National Programs to Support the Semiconductor Industry,* Charles W. Wessner, ed., Washington, D.C.: National Academies Press, 2003.

[43]Stephen S. Cohen and John Zysman, *Manufacturing Matters: The Myth of the Post-Industrial Economy*, New York: Basic Books, 1988.

[44]Among several bills introduced in Congress in the 2004 election year was that offered by Senators Kennedy and Daschle, which required that companies that send jobs abroad report how many, where, and why, giving 90 days notice to employees, state social service agencies, and the U.S. Labor Department. Senator John Kerry had also introduced legislation in 2004 requiring call center workers to identify the country they were phoning from.

Box E: Key Issues from the Participants' Roundtable

Given the understanding generated at the symposium about the uniqueness and complexity of software and the ecosystem that builds, maintains, and manages it, Dr. Raduchel asked each member of the final Participants' Roundtable to identify key policy issues that need to be pursued.

Drawing from the experience of the semiconductor industry, Dr. Flamm noted that it is best to look ahead to the future of the industry rather than look back and "invest in the things that our new competitors invest in," especially education. Dr. Rosing likewise pointed out the importance of lifelong learning, observing that the fact that many individuals did not stay current was a major problem facing the United States labor force. What the country needed, he said, was a system that created extraordinary incentives for people to take charge of their own careers and their own marketability.

Mr. Socas noted that the debate over software offshoring was not the same as a debate over the merits of free-trade, since factors that give one country a relative competitive advantage over another are no longer tied to a physical locus. Calling it the central policy question of our time, he wondered if models of international trade and system of national accounting, which are based on the idea of a nation-state, continue to be valid in a world where companies increasingly take a global perspective. The current policy issue, he concluded, concerns giving American workers the skills that allow them to continue to command high wages and opportunities. Also observing that the offshoring issue was not an ideological debate between free trade and protectionism, Dr. Hira observed that "we need to think about how to go about making software a viable profession and career for people in America."

What is required, in the final analysis, is a constructive policy approach rather than name calling, noted Dr. Hira. He pointed out that it was important to think through and debate all possible options concerning offshoring rather than tarring some with a "protectionist" or other unacceptable label and "squelching them before they come up for discussion." Progress on better data is needed if such constructive policy approaches are to be pursued.

PROGRESS ON BETTER DATA

Drawing the conference to a close, Dr. Jorgenson remarked that while the subject of measuring and sustaining the New Economy had been discovered by

"Wait a minute! We discovered this problem in 1999, and only five years later, we're getting the data."

Dale Jorgenson

economists only in 1999, much progress had already been made towards developing the knowledge and data needed to inform policy making. This conference, he noted, had advanced our understanding of the nature of software and the role it plays in the economy. It had also highlighted pathbreaking work by economists like Dr. Varian on the economics of open-source software, and Drs. Berndt and White on how to measure prepackaged software price while taking quality changes into account. Presentations by Mr. Beams and Ms. Luisi had also revealed that measurement issues concerning software installation, business reorganization, and process engineering had been thought through, with agreement on new accounting rules.

As Dr. Jorgenson further noted, the Bureau of Economic Analysis had led the way in developing new methodologies and was soon getting new survey data from the Census Bureau on how much software was being produced in the United States, how much was being imported, and how much the country was exporting. As national accountants around the world adopted these standards, international comparisons will be possible, he added, and we will be able to ascertain what is moving where—providing the missing link to the offshore outsourcing puzzle.

II

PROCEEDINGS

Introduction

Dale W. Jorgenson
Harvard University

Dr. Jorgenson, Chair of the National Research Council's Board on Science, Technology, and Economic Policy (STEP), welcomed participants to the day's workshop on Software Growth and the Future of the U.S. Economy. The program, whose agenda he acknowledged as very ambitious, was the fourth in the Board's series "Measuring and Sustaining the New Economy." The series was begun in the midst of a tremendous economic boom, and although conditions had changed, the basic structural factors had not: A new economy has in fact had momentous impact on productivity growth in the United States and around the world, and it is therefore of great importance for economic policy and for the country's future.[1]

[1]In the context of this analysis, the New Economy does not refer to the boom economy of the late 1990s. The term is used in this context to describe the acceleration in U.S. productivity growth that emerged in the mid-1990s, in part as a result of the acceleration of Moore's Law and the resulting expansion in the application of lower cost, higher performance information technologies. See Dale W. Jorgenson; Kevin J. Stiroh; Robert J. Gordon; Daniel E. Sichel, "Raising the Speed Limit. Raising the Speed Limit: U.S. Economic Growth in the Information Age," *Brookings Papers on Economic Activity*, (1):125-235, 2000.

PRODUCTIVITY AND MOORE'S LAW

Recapping the series to date, Dr. Jorgenson noted that it began by addressing a hardware phenomenon, the development of semiconductor technology, which he called "the most basic story of the New Economy." He used a graph to illustrate its driving force: the growth of capacity on memory chips and on logic chips, which are the basic hardware components for computers, and, increasingly, for communications equipment as well (See Figure 1). Translating this technical description into economic terms, he showed a graph of the relative prices of semiconductors from 1977 to 2000 (See Figure 2). Only in the previous 5 years or so had "relatively reasonable" data on semiconductor prices become available in the United States. But as information piled up, price changes for semiconductors were seen to mirror the dramatic developments of technology: Moore's Law,

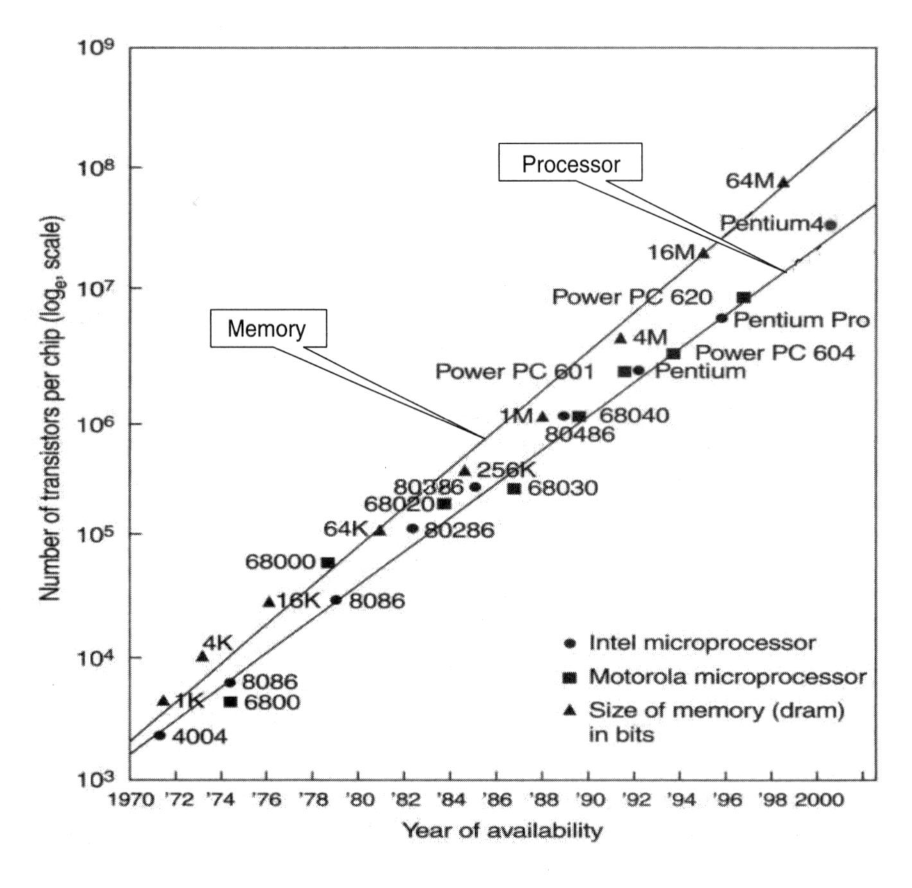

FIGURE 1 Transistor density on microprocessors and memory chips.

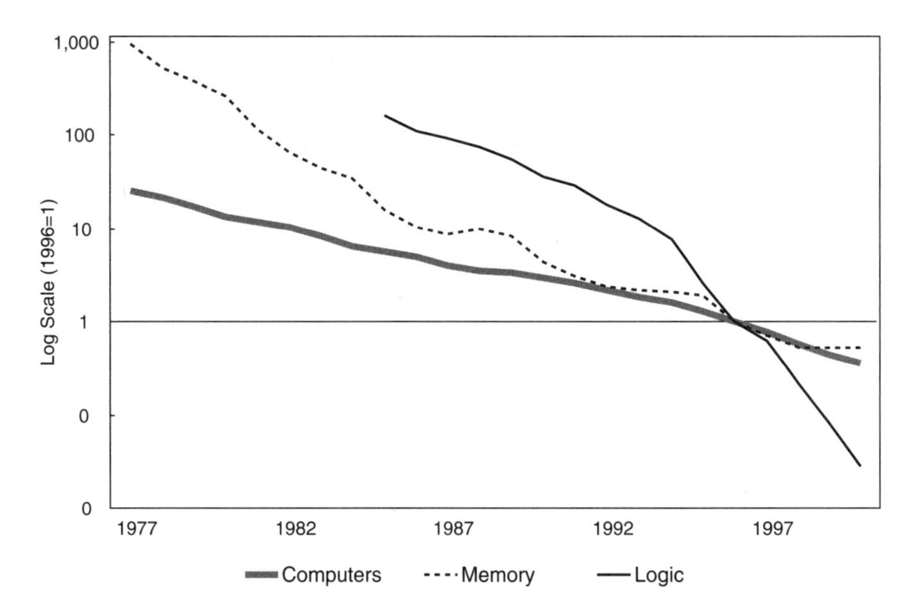

FIGURE 2 Relative prices of computers and semiconductors, 1977-2000.
NOTE: All price indexes are divided by the output price index.

which expresses a doubling of the number of transistors on a chip every 18 to 24 months, was mirrored in the price indexes for semiconductors.

Extending this information to such hardware as computers and communications equipment was the next step. Since 1985, the Bureau of Economic Analysis (BEA) of the U.S. Department of Commerce has maintained price indexes for computers in the national accounts. As a result of this "very satisfactory situation," economists have a clear idea of the obviously momentous effect that computers have on economic growth. Dr. Jorgenson held up the collaboration that produced these indexes, in which the government has been represented by BEA and the private sector by IBM, as a paradigm for STEP's program on Measuring and Sustaining the New Economy. He noted that the Board had designed its program to bring together those doing the measuring, who are mainly in government agencies, with those who know the technology, who are mainly in the private sector.

THREE CATEGORIES OF SOFTWARE

Within U.S. national accounts software is broken down into three categories: prepackaged, custom, and own-account. Prepackaged software, although just what it says—shrink-wrapped—is more commonly purchased with the computer itself; the software or patches to it are also often downloaded from a Web site. Custom software comes from firms like SAP or Oracle that produce large software systems to perform business functions—database management, human-resource management, cost accounting, and so on—and must be customized for the user. Own-account software refers to software systems built for a unique purpose, generally a large project such as managing a weapons system or an airline reservations system.

Dr. Jorgenson noted that quite a bit of price information has been gathered on prepackaged software, but that it makes up only 25 to 30 percent of the software market. Custom and own-account software are not as easy to measure as is shrink-wrapped, something reflected in the fact that price information on both of these components is scarcer than it is on prepackaged software and that no price index exists for either. As a consequence, "there is a large gap in our understanding of the New Economy," he said, adding that the aim of the day's workshop was to begin trying to fill this gap in.

A HISTORY OF DECLINING PRICES

Referring again to the chart, which he dubbed the "gold standard" for measurement issues, Dr. Jorgenson emphasized the tremendous declines in prices that it depicts (See Figure 2). While the chart goes back only as far as 1977, computer prices have declined at about 15 percent per year since the computer's commercialization. Semiconductor prices have been declining even more rapidly, about 50 percent a year for logic chips and about 40 percent a year for memory chips.

His next chart, based on a historic series constructed by BEA, showed prepackaged software prices declining at rates comparable to those of hardware as represented by computers and by communications equipment; prices of the latter, which relies increasingly on semiconductor technology, have behaved rather similarly to those of computers (See Figure 3). That software improvement, although not based on Moore's Law, has paralleled hardware's trend for a long time is "a bit of a mystery," he said.

Having imparted what he called "the good news," Dr. Jorgenson promised that William J. Raduchel, next to speak on "The Economics of Software," would "explain the bad news: that we don't have a very clear understanding collectively of the economics of software." Dr. Raduchel, Dr. Jorgenson's former colleague in economics at Harvard, designed early software for econometrics and model simulation. At this, Dr. Jorgenson remarked, he had such success that he left the academy for high-tech industry and ultimately, through a succession of steps, became chief technology officer of AOL-Time Warner.

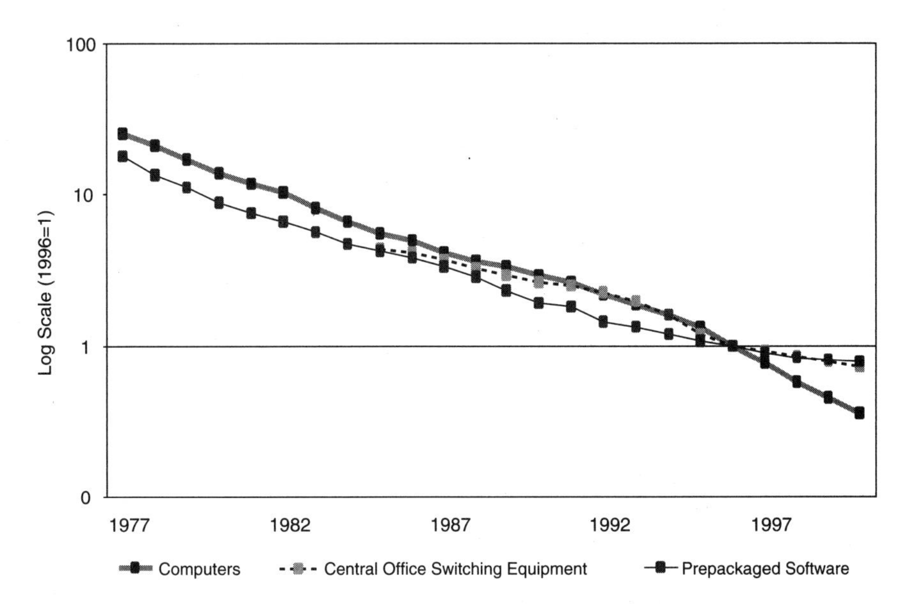

FIGURE 3 Relative prices of computers, communications, and software, 1977-2000.
NOTE: All price indexes are divided by the output price index.

The Economics of Software

William J. Raduchel

Dr. Raduchel noted that he was drawing from his many years of experience with software, which began when he wrote his first line of code as a teenager, and from his background as an economist to search for a good economic model for software. Yet, he cautioned that his presentation would not answer all the questions it raised; achieving a thorough understanding of the problem, he predicted, would take years, maybe even decades. Though sharing Dr. Jorgenson's concern about the consequences stemming from the lack of a practical economic model, however, he noted that he would attempt to bring the technology and the economics together in his presentation.

Dr. Raduchel characterized software as "the medium through which information technology [IT] expresses itself." Software loses all meaning in the absence of the computer, the data, or the business processes. Nonetheless, it is the piece of IT that is becoming not only increasingly central but also increasingly hard to create, maintain, and understand. Positing that software is the world's largest single class of either assets or liabilities, and the largest single class of corporate expenses, he argued that "the care and feeding of the systems [that] software runs...dominates the cost of everything." In addition, software is, as shown by the work of Dr. Jorgenson and a number of other economists, the single biggest driver of productivity growth.[2]

U.S. ECONOMY'S DEPENDENCE ON SOFTWARE

Projecting a satellite photograph of the United States taken during the electrical blackout of August 2003, Dr. Raduchel noted that a software bug had recently been revealed as the mishap's primary driver. When "one line of code buried in an energy management system from GE failed to work right," a vast physical infrastructure had been paralyzed, one example among many of how the U.S. economy "is so dependent in ways that we don't understand." Recalling the STEP Board's February 2003 workshop "Deconstructing the Computer," Dr. Raduchel described an incident related by a speaker from a company involved in systems integration consulting: When the credit card systems went down at a major New York bank, the chief programmer was brought in. After she had spent several minutes at the keyboard and everything had begun working again, "she pushed her chair slowly away from the desk and said, 'Don't touch anything.'

" 'What did you change?' the head of operations asked.

" 'Nothing,' she said. 'I don't know what happened.'

[2]Throughout the 1970s and 1980s, Americans and American businesses regularly invested in ever more powerful and cheaper computers, software, and communications equipment. They assumed that advances in information technology—by making more information available faster and cheaper—would yield higher productivity and lead to better business decisions. The expected benefits of these investments did not appear to materialize—at least in ways that were being measured. Even in the first half of the 1990s, productivity remained at historically low rates, as it had since 1973. This phenomenon was called "the computer paradox," after Robert Solow's casual but often repeated remark in 1987: "We see the computer age everywhere except in the productivity statistics." (See Robert M. Solow, "We'd Better Watch Out," *New York Times Book Review*, July 12, 1987.) Dale Jorgenson resolved this paradox, pointing to new data that showed that change at a fundamental level was taking place. While growth rates had not returned to those of the "golden age" of the U.S. economy in the 1960s, he noted that new data did reveal an acceleration of growth accompanying a transformation of economic activity. This shift in the rate of growth by the mid-1990s, he added, coincided with a sudden, substantial, and rapid decline in the prices of semiconductors and computers; the price decline abruptly accelerated from 15 to 28 percent annually after 1995. (See Dale W. Jorgenson and Kevin J. Stiroh, "Raising the Speed Limit: U.S. Economic Growth in the Information Age," in National Research Council, *Measuring and Sustaining the New Economy*, Dale W. Jorgenson and Charles W. Wessner, eds., Washington, D.C.: National Academies Press, 2002, Appendix A.) Relatedly, Paul David has argued that computer networks had to be sufficiently developed in order for IT productivity gains to be realized and recognized in the statistics. See Paul A. David, *Understanding the Digital Economy*, Cambridge, MA: MIT Press, 2000. Also see Erik Brynjolfsson and Lorin M. Hitt, "Computing Productivity: Firm-Level Evidence," *Review of Economics and Statistics* 85(4):793-808, 2003, where Brynjolfsson and Hitt argue that much of the benefit of IT comes in the form of improved product quality, time savings and convenience, which rarely show up in official macroeconomic data. Of course, as Dr. Raduchel noted at this symposium, software is necessary to take advantage of hardware capabilities.

"At that point the CEO, who was also in the room, lost it: He couldn't understand why his bank wasn't working and nobody knew why," Dr. Raduchel explained, adding: "Welcome to the world of software."

FROM BUBBLE TO OUTSOURCING: SPOTLIGHT ON SOFTWARE

In the 1990s bubble, software created a new and important part of the economy with millions of high-paying jobs. The spotlight has returned to software because it is emerging as a key way in which India and China are upgrading their economies: by moving such jobs out of the United States, where computer science continues to decline in popularity as a field of study. A discussion was scheduled for later in the program of "why it is in the United States' best interest to tell this small pool of really bright software developers not to come here to work" by denying them visas, with the consequence that other jobs move offshore as well. Although the merits can be argued either way, Dr. Raduchel said, to those who are "worried about the country as a whole . . . it's becoming a major issue."

In the meantime, hardware trends have continued unchanged. Experts on computer components and peripherals speaking at "Deconstructing the Computer" predicted across-the-board price declines of 20 to 50 percent for at least 5 more years and, in most cases, for 10 to 20 years.[3] During this long period marked by cost reductions, new business practices will become cost effective and capable of implementation through ever more sophisticated software. As a result, stated Dr. Raduchel, no industry will be "safe from reengineering or the introduction of new competitors."

Characterizing a business as "an information system made up of highly decentralized computing by fallible agents, called people, with uncertain data," Dr. Raduchel asserted that there is no difference to be found "at some level of reduction . . . between economics and information technology." Furthermore, in many cases, a significant part of the value of a firm today is tied up in the value of its software systems.[4] But, using the firm Google as an example, he pointed out that its key assets, its algorithms do not show up on its balance sheet.[5] "Software

[3]This refers to the price of equivalent functionality. In fact, functionality tends to increase over time, making the actual price declines less dramatic. In addition, this prediction is just for the semiconductor content; prices of finished goods can be expected to decline much less. For example, the decline in price of a typical desktop computer has been far less dramatic than predicted by Moore's Law. The computer includes many costs other than semiconductors such as boards, assembly, packaged software, marketing and sales, and overhead.

[4]Given that software is a key medium of interaction within a modern firm, much of the tacit knowledge held by management and employees is meshed in with the software. In addition, a modern firm's algorithms are increasingly embedded and executed through its specialized and non-specialized software.

[5]Accounting standards have wrestled with the problems of capitalization of intangible assets like software, and it remains a highly controversial issue in the accounting profession.

is intangible and hard to measure, so we tend not to report on it," he observed, "but the effect is mainly to put people in the dark."

REDEFINING THE CONSUMER'S WORLD

Software is totally redefining the consumer's world as well. There are now scores of computers in a modern car, and each needs software, which is what the consumer directly or indirectly interacts with. The personal computer, with devices like iPod that connect to it, has become the world's number-one device for playing and managing music, something that only 5 years before had been a mere speck on the horizon. With video moving quickly in the same direction, Dr. Raduchel predicted, the next 10 years would be wrenching for all consumer entertainment, and piracy would be a recurrent issue. A major factor driving piracy, he noted is that "the entertainment industry makes money by not delivering content in the way consumers want." Music piracy began because consumers, as has been clear for 15 years, want music organized with a mix of tracks onto a single play-list, but that doesn't suit any of the music industry's other business models.

Anticipating an aspect of Monica Lam's upcoming talk, Dr. Raduchel observed that software may look "remarkably easy and straightforward," but the appearance is borne out in reality only to a certain extent. Among those in the audience at one of two Castle Lectures he had given the previous year at the U.S. Military Academy were numerous freshmen who had recently written programs 20 to 30 lines in length and "thought they really understood" software. But a program of that length, even if some have difficulty with it, is "pretty easy to write." The true challenge, far from creating a limited piece of code, is figuring out how to produce software that is "absolutely error-free, robust against change, and capable of scaling reliably to incredibly high volumes while integrating seamlessly and reliably to many other software systems in real time."

SOFTWARE: IT'S THE STACK THAT COUNTS

For what matters, rather than the individual elements of software, is the entirety of what is referred to as the *stack*. The software stack, which comprises hundreds of millions of lines of code and is what actually runs the machine, begins with the *kernel*, a small piece of code that talks to and manages the hardware. The kernel is usually included in the operating system, which provides the basic services and to which all programs are written. Above the operating system is middleware, which hides both the operating system and the window manager, the latter being what the user sees, with its capacity for creating windows, its help functions, and other features. The operating system runs other programs called *services*, as well as the applications, of which Microsoft Word and PowerPoint are examples.

"When something goes right or goes wrong with a computer, it's the entire software stack which operates," stated Dr. Raduchel, emphasizing its complexity. The failure of any given piece to work when added to the stack may not indicate that something is wrong with that piece; rather, the failure may have resulted from the inability of some other piece of the stack, which in effect is being tested for the first time, to work correctly with the new addition. A piece of packaged software is but one component of the stack, he said, and no one uses it on its own: "The thing you actually use is the entire stack."

Finally, the software stack is defined not only by its specifications, but also by the embedded errors that even the best software contains, as well as by undocumented features. To build a viable browser, a developer would have to match Internet Explorer, as the saying goes, "bug for bug": Unless every error in Internet Explorer were repeated exactly, the new browser could not be sold, Dr. Raduchel explained, because "other people build to those errors." While data are unavailable, an estimate that he endorsed places defects injected into software by experienced engineers at one every nine or ten lines.[6] "You've got a hundred million lines of code?" he declared. "You do the arithmetic."

With hundreds of millions of lines of code making up the applications that run a big company, and those applications resting on middleware and operating systems that in turn comprise tens of millions of lines of code, the average corporate IT system today is far more complicated than that of the Space Shuttle or Apollo Program. And the costs of maintaining and modifying software only increase over time, since modifying it introduces complexity that makes it increasingly difficult to change. "Eventually it's so complicated," Dr. Raduchel stated, "that you can't change it anymore. No other major item is as confusing, unpredictable, or unreliable" as the software that runs personal computers.

THE KNOWLEDGE STACK OF BUSINESS PROCESSES

Opposite the software stack, on the business side, is what Dr. Raduchel called the *knowledge stack*, crowned by the applications knowledge of how a business actually runs. He ventured that most of the world's large organizations would be unable to re-implement the software they use to run their systems today, in the absence of the skilled professionals whose knowledge is embedded in them. "Stories are legion in the industry about tracking down somebody at a retirement home in Florida and saying, 'Here's $75,000. Now, would you please tell me what you did 10 years ago, because we can't figure it out?' " Systems knowledge is the ability to create a working system that operates the applications at the top of

---

[6]Real defect insertion rates are difficult to measure, because it depends on when the measurements are made. Some experts suggest that most defects are caught by developers within a few minutes of their being inserted; other are caught the first time they attempt to compile code, maybe a few hours later.

the stack; computer knowledge amounts to truly understanding what the system does at some level, and especially at the network level. Because almost no individual has all this knowledge, systems development is a team activity unique in the world.

The industry has dealt with this real-world challenge over the past 30 years by making the software stack more abstract as one moves up so that, ideally, more people are able to write software. Those who can write at the kernel level of the stack number in the hundreds at best worldwide. For, as is known to those with experience in the software industry, the very best software developers are orders of magnitude better than the average software developer. "Not 50 percent better, not 30 percent, but 10 times, 20 times, 100 times better," Dr. Raduchel emphasized. So a disproportionate amount of the field's creative work is done by "a handful of people: 80, 100, 200 worldwide."[7] But millions of people can write applications in Microsoft Visual Basic; the results may not be very good in software terms, but they may be very useful and valuable to those who write them. The rationale for introducing abstraction is, therefore, that it increases the number who can write and, ideally, test software. But, since heightening abstraction means lowering efficiency, it involves a trade-off against computing power.

THE MAJOR COSTS: CONFIGURATION, TESTING, TUNING

The major costs involved in making a system operational are configuration, testing, and tuning. Based on his experience advising and participating in "over a hundred corporate reengineering projects," Dr. Raduchel described the overall process as "very messy." Packaged software, whose cost can be fairly accurately tracked, has never represented more than 5 percent of the total project cost; "the other 95 percent we don't track at all." Accounting rules that, for the most part, require charging custom and own-account software to general and administrative (G&A) expense on an ongoing basis add to the difficulty of measurement.

A lot of labor is needed, with 1 designer to 10 coders to 100 testers representing a "good ratio." Configuration, testing, and tuning account for probably 95 to 99 percent of the cost of all software in operation. Recalling Dr. Jorgenson's allusion to the downloading of patches, Dr. Raduchel noted that doing so can force changes. That, he quipped, is "what makes it so much fun when you run Microsoft Windows Update: [Since] you can change something down here and break something up here . . . you don't know what's going to not work after you've done it." Moreover, because of the way software ends up being built, there's no way around such trade-offs.

[7]Estimates of this ratio vary within the industry. Some believe that test and evaluation costs are half the total development cost, not 95-99 percent. In many cases, requirements for engineering, architecture, and design account for large portions of the overall cost of software development.

ECONOMIC MODELS OF SOFTWARE OUT OF DATE

As a result, software is often "miscast" by economists. Many of their models, which treat software as a machine, date to 40 years ago, when software was a minor portion of the total cost of a computer system and was given away. The economist's problem is that software is not a factor of production like capital and labor, but actually embodies the production function, for which no good measurement system exists. "Software is fundamentally a tool without value, but the systems it creates are invaluable," Dr. Raduchel stated. "So, from an economist's point of view, it's pretty hard to get at—a black box." Those who do understand "the black arts of software" he characterized as "often enigmatic, unusual, even difficult people"—which, he acknowledged, was "probably a self-description."

Producing good software, like producing fine wine, requires time. IBM mainframes today run for years without failure, but their software, having run for 30 years, has in effect had 30 years of testing. Fred Brooks, who built the IBM system, taught the industry that the only way to design a system is to build it. "Managers don't like that because it appears wasteful," Dr. Raduchel said, "but, believe me, that's the right answer." For specifications are useless, he said, noting that he had never seen a complete set of specifications, which in any case would be impractically large. Full specs that had recently been published for the Java 2 Enterprise Edition, the standard for building mini-corporate information systems, are slightly more than 1 meter thick. "Now, what human being is going to read a meter of paper and understand all its details and interactions?" he asked, adding that the content was created not by a single person but by a team.

Their inability to measure elements of such complexity causes economists numerous problems. A good deal of software is accounted for as a period expense; packaged software is put on the balance sheet and amortized. While programming is what people usually think of as software, it rarely accounts for more than 10 percent of the total cost of building a system. The system's design itself is an intangible asset whose value grows and shrinks with that of the business it supports. The implementation also has value because it is often very difficult to find even one implementation that will actually work. Despite all this, Dr. Raduchel remarked, only one major corporation he knew of recognized on its books the fact that every system running must be replaced over time.

MAJOR PUBLIC POLICY QUESTIONS

Dr. Raduchel then turned to public-policy questions rooted in the nature and importance of software:

- **Are we investing adequately in the systems that improve productivity?** Numerous reports have claimed that the enterprise resource planning and other systems put into place in the late 1990s in anticipation of the year 2000 have greatly improved the efficiency of the economy, boosting productivity and contributing to low inflation. But many questions remain unanswered: How much did we invest? How much are we investing now, and is that amount going up or down? How much should we be investing? What public-policy measures can we take to encourage more investment? There are no data on this, but anecdotal evidence suggests that investment has fallen significantly from its levels of 7 years ago. Companies have become fatigued and in many cases have fired the chief information officers who made these investments. "People are in maintenance mode," said Dr. Raduchel. Noting that systems increasingly become out of synch with needs as needs change over time, he warned that "systems tend to last 7 years, give or take; 7 years is coming up, and some of these systems are going to quit working and need to be replaced."[8]

• **Do public corporations properly report their investments and the resulting expenses?** And a corollary: How can an investor know the worth of a corporation that is very dependent on systems, given the importance of software to the value of the enterprise and its future performance? Mining the history of the telecommunications industry for an example, Dr. Raduchel asserted that the billing system is crucial to a company's value, whereas operating a network is "pretty easy." Many of the operational problems of MCI WorldCom, he stated, arose "from one fact and one fact only: It has 7 incompatible billing systems that don't talk to one another." And although the billing system is a major issue, it is not covered in financial reports, with the possible exceptions of the management's discussion and analysis (MD&A) in the 10K.

• **Do traditional public policies on competition work when applied to software-related industries?** It is not clear, for example, that antitrust policies apply to software, which develops so rapidly that issues in the industry have changed before traditional public policy can come into play. This "big question" was being tested in *United States v. Microsoft.*

• **Do we educate properly given the current and growing importance of software?** What should the educated person know about software? Is sufficient training available? Dr. Raduchel noted that the real meat of the Sarbanes-Oxley Act is control systems rather than accounting.[9] "I chair an audit committee, and I

[8]There is also the issue of the legacy stack on which the software runs, the support and upgrade of which over time will be abandoned by its vendors.

[9]The Sarbanes-Oxley Act of 2002 was passed to restore the public's confidence in corporate governance by, among other measures, making chief executives of publicly traded companies personally validate financial statements and other information. Maintaining such proper "internal controls," however, requires secure computer systems for maintaining data. Thus, while Sarbanes-Oxley doesn't mandate specific internal controls such as strong authentication or the use of encryption, the law has persuaded corporate executives of the need to ensure that proper and auditable security measures are in place, as they could face criminal penalties if a breach is detected.

love my auditor, but he doesn't know anything about software," he said. And if it took experts nearly a year to find the bug that set off the 2003 blackout, how are lay people to understand software-based problems?

- **What should our policy be on software and business-methods patents?** Those long active in the software industry rarely see anything patented that had not been invented 30 years ago; many patents are granted because the prior art is not readily available. Slashdot.org, an online forum popular in the programming community, has a weekly contest for the most egregious patent. "If you read the 10Ks, people are talking about [these patents'] enormous value to the company," Dr. Raduchel said. "Then you read Slashdot and you see that 30 pieces of prior art apply to it." The European Union is debating this issue as well.

- **What is the proper level of security for public and private systems, and how is it to be achieved?** A proposal circulating in some federal agencies would require companies to certify in their public reporting the security level of all their key systems, particularly those that are part of the nation's critical infrastructure. Discussed in a report on security by the President's Council of Advisors on Science and Technology (PCAST) was the amount of risk to the economy from vulnerability in systems that people may never even have thought about. The worms that had been circulating lately, some of which had caused damage in the billions of dollars, were so worrisome because they could be used to crack computers in 10 million to 20 million broadband-connected homes and to create an attack on vital infrastructure that would be "just unstoppable," said Dr. Raduchel, adding: "Unfortunately, this is not science fiction but a real-world threat. What are we going to do about it?"

- **What is happening to software jobs?** Do we care about their migration to India and China? Is U.S. industry losing out to lower-cost labor abroad or are these jobs at the very tip of the value chain, whose departure would make other parts of high-tech industry hard to sustain in the United States? "Let me tell you," Dr. Raduchel cautioned, "the people in India and China think it's really important to get those software jobs."

- **What export controls make sense for software?** Taken at the margin, all software has the potential for dual use. As Dr. Raduchel noted wryly, "If you're building weapons for Saddam Hussein, you still have to make Powerpoint presentations to him about what's going on—apparently they were all full of lies, but that's another issue." More practically, however, export controls for dual-use software, such as those calling for encryption, can help ensure that certain types of sensitive software are not used in a way that is detrimental to U.S. national security.

• **Should the building of source code by a public community—open-source code—be encouraged or stopped?** Dr. Raduchel included himself among those who have come to believe that source code is the least valuable rather than the most valuable part of software. Consequently, giving it away is actually a good strategy. However, some forces in the United States, primarily the vendors of proprietary software, want to shut open-source software down, while others, such as IBM have played an important role in developing Linux and other open-source platforms.[10] In the course of a discussion with Federal Communications Commission Chairman Michael Powell, Dr. Raduchel recounted that he had suggested that all critical systems should be based on open-source software because it is more reliable and secure than proprietary software. Some believe that if nobody knows what the software is, it would be more reliable and secure. Yet, that position overlooks the view that open-source software is likely to have fewer bugs.[11] Many, according to Dr. Raduchel, "would argue that open-source software is going to be inherently more reliable and secure because everybody gets to look at it."

• **What liability should apply to sales of software?** Licenses on shrink-wrapped software specify "software sold as is"; that is, no warranty at all is provided. In view of the amount of liability being created, should that be changed? And, if so, what should be done differently? Dr. Raduchel called the central problem here the fact that, although a new way of writing software is probably needed none has emerged, despite much research a couple of decades back and ceaseless individual effort. Bill Joy, one of the founders of BSD UNIX, had recently stated that all methods were antique and urged that a new way be found, something Dr. Raduchel rated "hugely important" as the potential driver of "a value equation that is incredibly powerful for the country."

• **How are we investing in the technology that creates, manages, and builds software?** Outside of the National Science Foundation and some other institutions that are stepping up funding for it, where is the research on it?

Pointing to the richness of this public-policy agenda, Dr. Raduchel stated: "I am not sure we are going to get to any answers today; in fact, I am sure we're

---

[10]In some cases, larger firms such as IBM and Sun have played an important role in developing open-source software platforms such as Linux. See Jan Stafford, "LinuxWorld: IBM Stimulates Development of Linux Apps," SearchEnterpriseLinux.com, August 2, 2004.

[11]Expert opinions differ on the relative reliability and security of open-source software vis-à-vis proprietary software. Essentially, it is a measurement issue turning on how reliability is measured. For an analysis concerning the measurement of quality, see T. J. Halloran and William Scherlis, "High Quality and Open Source Software Practices," Position Paper, 24th International Conference on Software Engineering, 2002.

not." But he described the day's goal for the STEP Board, including Dr. Jorgenson and himself, and for its staff as getting these issues onto the table, beginning a discussion of them, and gaining an understanding of where to proceed. With that, he thanked the audience and turned the podium back over to Dr. Jorgenson.

Panel I

The Role of Software—What Does Software Do?

INTRODUCTION

Dale W. Jorgenson
Harvard University

Calling Dr. Raduchel's presentation a "brilliant overview" of the terrain to be covered during the day's workshop, Dr. Jorgenson explained that the two morning sessions would be devoted to technology and that the afternoon sessions would focus on economics, in particular on measurement problems and public-policy questions relating to software. He then introduced Tony Scott, the Chief Information Technology Officer of General Motors.

Anthony Scott
General Motors

Dr. Scott responded to the question framing the session—"What does software do?"—by saying that, in the modern corporation, software does "everything." Seconding Dr. Raduchel's characterization of software as the reduction and institutionalization of the whole of a corporation's knowledge into business processes and methods, he stated: "Virtually everything we do at General Motors has been reduced in some fashion or another to software." Many of GM's products have become reliant on software to the point that they could not be sold, used, or serviced without it.

By way of illustration, Dr. Scott noted that today's typical high-end automobile may have 65 microprocessors in it controlling everything from the way fuel is used to controlling airbags and other safety systems. Reflecting how much the automobile already depends on software, internal GM estimates put the cost of the electronics, including software, well above that of the material components—such as steel, aluminum, and glass—that go into a car. This trend is expected to continue. As a second example he cited what he called "one of [GM's] fastest-growing areas," "OnStar," a product that, at the push of a button, provides drivers a number of services, among them a safety and security service. If a car crashes and its airbags deploy, an OnStar center is notified automatically and dispatches emergency personnel to the car's location. "That is entirely enabled by software," he said. "We couldn't deliver that service without it."

To further underscore the importance of software to the automobile business, Dr. Scott turned to the economics of leasing. When a car's lease is up, that car is generally returned to the dealer, who then must make some disposition of it. In the past, the car would have been shipped to an auction yard, where dealers and others interested in off-lease vehicles could bid on it; from there, it might be shipped two or three more times before landing with a new owner. But under "Smart Auction," which has replaced this inefficient procedure, upon return to the dealer a leased car is photographed; the photo is posted, along with the car's vital statistics, on an auction Web site resembling a specialized eBay; and the car, sold directly to a dealer who wants it, is shipped only once, saving an average of around $500 per car in transportation costs. "The effects on the auto industry have been enormous," Dr. Scott said. "It's not just the $500 savings; it's also the ability of a particular dealer to fine-tune on a continuous basis the exact inventory he or she wants on the lot without having to wait for a physical auction to take place." This is just one of many examples, he stressed, of software's enabling business-process changes and increased productivity across every aspect of his industry.

Downside of Software's Impact

But Dr. Scott acknowledged that there is also a "bad-news side" to software's impact and, to explain it, turned to the history of Silicon Valley's famed Winchester Mystery House, which he called his "favorite analogy to the software business." Sarah Winchester, who at a fairly young age inherited the Winchester rifle fortune, came to believe that she must keep adding to her home constantly, for as long as it was under construction, she would not die. Over a period of 40 or 50 years she continually built onto the house, which has hundreds of rooms, stairways that go nowhere, fireplaces without chimneys, doors that open into blank walls. It is very finely built by the best craftsmen that money could buy, and she paid good architects to design its various pieces. But, taken as a whole, the house is unlivable. "It has made a fairly good tourist attraction over the years, and I urge you to go visit it if you ever visit Silicon Valley, because it is a marvel," Dr. Scott said. "But it's unworkable and unsustainable as a living residence."

The process by which corporations build software is "somewhat analogous to the Winchester Mystery House," he declared, citing an " 'add-onto effect': I have a bunch of systems in a corporation and I'm going to add a few things on here, a few things on there, I'm going to build a story there, a stairway, a connector here, a connector there." Over time, he said, harking back to Dr. Raduchel's remarks, the aggregation of software that emerges is so complex that the thought of changing it is daunting.

Sourcing and Costing Information Technology

For a case in point, Dr. Scott turned to the history of General Motors, which in the 1980s purchased EDS, one of the major computer-services outsourcing companies at the time. EDS played the role of GM's internal information technology (IT) organization, in addition to continuing to outsource for other companies, for a dozen or so years ending in 1996; then, GM spun EDS off as a separate company again, at the same time entering into a 10-year agreement under which EDS continued as its primary outsourcer. Owing to the lack of a suitable governance structure—which Dr. Scott called "the only flaw in the model"—GM managers with budget authority were, during the period when EDS was under GM's roof, basically allowed to buy whatever IT they needed or wanted for their division or department. The view within GM, as he characterized it, was: " 'Well, that's o.k., because all the profits stay in the company, so it's sort of funny money—not real money—that's being spent.' "

This behavior resulted in tremendous overlap and waste, which were discovered only after GM had spun EDS off and formed a separate organization to manage its own IT assets. Having developed "no economies of scale, no ability to leverage our buying power or standardize our business processes across the corporation," Dr. Scott recalled, GM had ended up with "one of everything that had ever been produced by the IT industry." One of the corporation's major efforts of the past 7 years had been taking cost out.

To illuminate the experience—which he doubted is unique to GM—Dr. Scott offered some data. GM started in 1996 with 7,000 applications it considered "business critical": applications whose failure or prolonged disruption would be of concern from a financial or an audit perspective. The corporation set in place objectives aimed at greatly reducing the number of systems in use "by going more common and more global with those systems" across the company. By early 2004, GM had reduced its systems by more than half, to a little over 2,000, and had in the process driven reliability up dramatically.

GM's IT Spending: From Industry's Highest to Lowest

But the "critical factor," according to Dr. Scott, is that GM's annual spending on information technology had dropped from over $4 billion in 1996 to a projected $2.8 billion in 2004—even though, in the interim, its sales had increased every year. This change in cost, he noted, could be measured very accurately because GM is 100 percent outsourced in IT, employing no internal staff to develop code, support systems, maintain systems, operate data centers, run networks, or perform any other IT function. It has meant that GM, which in 1996 had the highest percentage of IT spending in the automotive business, today has arguably the lowest, and with improved functionality and higher reliability. This "huge swing," which he described as underscoring "some of the opportunity" that exists to lower IT costs, had given "a little over $1 billion a year straight back to the company to invest in other things—like developing new automobiles with more software in them."

Addressing the issues of software quality and complexity, Dr. Scott endorsed Dr. Raduchel's depiction of the latter, commenting that the "incredible" complexity involved in so simple a task as delivering a Web page to a computer screen has been overcome by raising the level of abstraction at which code can be written. Because of the high number of microprocessors in the modern vehicle, the automotive sector is also struggling with complexity, but there is a further dimension: Since it must support the electronics in its products for a long time, the quality of its software takes on added importance. "Cars typically are on the road 10 years or more," he noted. "Now, go try to find support for 10-year-old software—or last year's software—for your PC."

Seeking a Yardstick for Software

Complicating the production of software good enough to be supported over time is the fact that, just like economic measures, such qualitative tools as standards and measures are lacking. "There are lots of different ways of measuring a piece of steel or aluminum or glass: for quality, for cost, and all the rest of it," Dr. Scott said. "But what is the yardstick for software?" There are no adequate measures either for cost or quality, the two most important things for industry to measure.

Dr. Scott then invited questions.

DISCUSSION

Dr. Jorgenson, admonishing members of the audience to keep their comments brief and to make sure they ended with question marks, called for questions to either Dr. Scott or Dr. Raduchel.

The Utility and Relevance of the CMM

Asked about the utility of the Capability Maturity Model (CMM) developed at Carnegie Mellon's Software Engineering Institute (SEI), Dr. Scott acknowledged the value of such quality processes as it and Six Sigma but noted that they have "not been reduced to a level that everyday folks can use." Recalling Dr. Raduchel's reference to the Sarbanes-Oxley Act, he said that much of what concerns corporations lies in questions about the adequacy of controls: Does the billing system accurately reflect the business activity that took place? Do the engineering systems accurately provide warning for issues that might be coming up in the design of products or services? Registering his own surprise that anyone at all is able to answer some of these questions, he said that he personally "would be very uncomfortable signing Sarbanes-Oxley statements in those areas."

Dr. Raduchel identified the stack issue as one of the challenges for CMM, observing that even if a software module is written according to the world's best engineering discipline, it must then go out into the real world. He sketched the dilemma before an IT officer faced with a quick decision on whether to apply a critical security patch—that, obviously, has never been tested for the company's application—to an operating system that has been performing well. On one side is the risk of leaving the system vulnerable to a security breach by trying to ensure it keeps going, on the other the risk of causing a functioning system to go down by trying to protect it. "There's no Carnegie Mellon methodology for how you integrate a hundred-million-line application system that's composed of 30 modules built over 20 years by different managers in different countries," he commented. The limits of the modular approach are often highlighted among the cognoscenti by comparing the construction of software to that of cathedrals: One can learn to build a particular nave, and to build it perfectly, but its perfection is no guarantee that the overall structure of which it is a part will stand.

Functioning in the Real World

To emphasize the point, Dr. Raduchel described an incident that occurred on the day the U.S. Distant Early Warning (DEW) Line went into operation in the late 1950s. Almost immediately, the system warned that the Soviet Union had launched a missile attack on the United States. Confronted with the warning, a major at NORAD in Colorado Springs wondered to himself: " 'The Soviets knew we were building this. They knew it was going live today. Why would they pick today to attack?' " He aborted the system's response mechanism to allow verification of the warning, which turned out to have been triggered by the rising of the moon over Siberia. "The fact was, the software didn't malfunction—it worked perfectly," Dr. Raduchel said. "But if the specs are wrong, the software's going to be wrong; if the stack in which it works has a bug, then it may not work right."

In the real world, the performance of a given application is a minor challenge compared to the operation of the full software stack. In Dr. Raduchel's experience of implementing major software projects, the biggest challenge arose in the actual operating environment, which could not be simulated in advance: "You'd put it together and were running it at scale—and suddenly, for the first time, you had 100 million transactions an hour against it. Just a whole different world."

In the Aftermath of Y2K

Bill Long of Business Performance Research Associates then asked what, if anything, had been learned from the Y2K experience, which refers to the widespread software fixes necessary to repair the inability of older computer systems to handle dates correctly after the millennium changes.

Responding, Dr. Raduchel called Y2K an "overblown event" and asserted that fear of legal liability, which he blamed the press for stoking, had resulted in billions of dollars of unnecessary spending. The measures taken to forestall problems were in many cases not well engineered, a cause for concern about the future of the systems they affect. Although Y2K had the positive result of awakening many to how dependent they are on software systems, it also raised the ire of CEOs, who saw the money that went to IT as wasted. "Most of the CIOs [chief information officers] who presided over the Year 2000 were in fact fired," he observed, and in "a very short period of time." As to the specifics of whether anything useful might have been learned regarding software investment, reinvestment, and replacement, Dr. Raduchel demurred, saying he had never seen any data on the matter. His general judgment, however, was that the environment was too "panicked" to make for a "good learning experience."

Dr. Scott, while agreeing with Dr. Raduchel's description of the negatives involved, sees the overall experience in a more positive light. Although a small number of system breakdowns did occur, the "whole ecosystem" proved itself sufficiently robust to give the lie to doomsday scenarios. In his personal experience of pre-Y2K testing, at the company where he worked prior to GM, some potential date-related problems were identified; they were corrected, but the consensus was that, even had they gone undetected, recovering from them would have been "fairly easy." The testing had a side benefit: It also led to the discovery of problems unconnected to Y2K. As for the downside, he noted that the CEOs' wrath and the CIOs' loss of credibility had been factors in the ensuing downturn in the IT economy.

How Standards Come into Play

The speakers were then asked to state their views on the economics of standards with respect to software, and in particular on how standards relate to the scalability of systems; on their impact on productivity at the macro level; and on their role in software interoperability.

Speaking as an economist, Dr. Raduchel lauded standards for opening up and expanding markets, and he declared that, in fact, "everything about them is good." But he cautioned that unless a software standard is accompanied by what a computer scientist would call a "working reference implementation," it is incomplete and of such limited value that, in the end, the volume leader defines the standard. He stressed that there is no way to write down a set of code completely as a specification, and one finds out what is left out only in the writing. In many cases where standards do exist, "the way Microsoft implements it is all that matters"—and should Microsoft choose not to follow the standard at all, any software product built to be interoperable with its products must follow Microsoft's bugs and whatever deliberate decisions it made not to be compatible. Foreshadowing Hal Varian's talk on open-source software, he noted that Linux provides one of the first instances in which both a powerful standard and a working reference implementation have appeared at the same time, and he credited that for Linux's emerging influence.

Dr. Jorgenson thanked Dr. Scott for conveying how software problems play out in the real world and lauded both speakers for providing a very stimulating discussion.

How Do We Make Software and Why Is It Unique?

INTRODUCTION

Dale W. Jorgenson
Harvard University

Dr. Jorgenson introduced the leadoff speaker of the second panel, Dr. Monica Lam, a professor of computer science at Stanford who was trained at Carnegie Mellon and whom he described as one of the world's leading authorities on the subject she was to address.

HOW DO WE MAKE IT?

Monica Lam
Stanford University

Dr. Lam commenced by defining software as an encapsulation of knowledge. Distinguishing software from books, she noted that the former captures knowledge in an executable form, which allows its repeated and automatic application to new inputs and thereby makes it very powerful. Software engineering is like many other kinds of engineering in that it confronts many difficult problems, but a number of characteristics nonetheless make software engineering unique: abstraction, complexity, malleability, and an inherent trade-off of correctness and function against time.

That software engineering deals with an abstract world differentiates it from chemical engineering, electrical engineering, and mechanical engineering, all of which run up against physical constraints. Students of the latter disciplines spend a great deal of time learning the laws of physics; yet there are tools for working with those laws that, although hard to develop initially, can be applied by many people in the building of many different artifacts. In software, by contrast, everything is possible, Dr. Lam said: "We don't have these physical laws, and it is up to us to figure out what the logical structure is and to develop an abstract toolbox for each domain." In Dr. Lam's opinion, this abstraction, which carries with it the necessity of studying each separate domain in order to learn how artifacts may be built in it, makes software much more difficult than other fields of engineering.

The Most Complex Thing Humans Build

As to complexity, a subject raised by previous speakers, software may be the most complex thing that humans have learned how to build. This complexity is at least in part responsible for the fact that software development is still thriving in the United States. Memory chips, a product of electrical engineering, have left this country for Japan, Korea, and Taiwan. But complexity makes software engineering hard, and the United States continues to do very well in it.

The third characteristic is malleability. Building software is very different from building many other things—bridges, for example. The knowledge of how to build bridges accrues, but with each new bridge the engineer has a chance to start all over again, to put in all the new knowledge. When it comes to software, however, it takes time for hundreds of millions of lines of code to accrue, which is precisely how software's complexity arises. A related problem is that of migrating the software that existed a long time ago to the present, with the requirements having changed in the meantime—a problem that again suggests an analogy to the Winchester Mystery House.

Trading Off Correctness Against Other Pluses

That no one really knows how to build perfect software—because it is abstract, complex, and at the same time malleable—gives rise to its fourth unique aspect: It can work pretty well even if it is not 100 percent correct.[12] "There is a choice here in trading off correctness for more features, more function, and better

[12]While this is true for software designed for "robustness" or "fault tolerance," in the absence of redundancy precision may be lost in results when systems are designed this way.

time to market," Dr. Lam explained, "and the question 'How do we draw this balance?' is another unique issue seen in software development."

Turning to the problem of measuring complexity, Dr. Lam noted that a theoretical method put forth by Kolmogorov measures the complexity of an object by the size of the Turing machine that generates it. Although it is the simplest kind of machine, the Turing machine has been proven equivalent in power to any supercomputer that can possibly be built. It does not take that many lines of Turing machine code to describe a hardware design, even one involving a billion transistors, because for the most part the same rectangle is repeated over and over again. But software involves millions and millions of lines of code, which probably can be compressed, but only to a limited degree. Pointing to another way of looking at complexity, she stated that everything engineered today—airplanes, for example—can be described in a computer. And not only can the artifact be described, it can be simulated in its environment.

The Stages of Software Development

Furthermore, the software development process comprises various stages, and those stages interact; it is a misconception that developers just start from the top and go down until the final stage is reached. The process begins with some idea of requirements and, beyond that, Dr. Lam said, "it's a matter of how you put it together." There are two aspects of high-level design—software architecture and algorithms—and the design depends on the kind of problem to be solved. If the problem is to come up with a new search engine, what that algorithm should be is a big question. Even when, as if often the case, the software entails no hard algorithmic problems, the problem of software architecture—the struggle to put everything together—remains. Once the high-level design is established, concern shifts to coding and testing, which is often carried out concurrently. These phases are followed by that of software maintenance, which she characterized as "a long tail."

Returning to requirements, she noted that these evolve and that it is impossible to come up all at once with the full stack. Specifying everything about how a piece of software is supposed to work would yield a program not appreciably smaller than the voluminous assemblages of code that now exist. And the end user does not really know what he or she wants; because there are so many details involved, the only way to identify all the requirements is to build a working prototype. Moreover, requirements change across time and, as software lasts awhile, issues of compatibility with legacy systems can arise.

As for architectural design, each domain has its own set of concepts that help build systems in the domain, and the various processes available have to be automated. Just as in the case of requirements, to understand how to come up with the right architecture, one must build it. Recalling the injunction in *The Mythical Man-Month,* Fred Brook's influential book, to "throw one away" when building a system, Dr. Lam said that Brooks had told her some months before: " 'Plan to throw one away' is not quite right. You have to plan to throw several of them away because you don't really know what you are doing; it's an iterative process; and you have to work with a working system."[13] Another important fact experience has taught is that a committee cannot build a new architecture; what is needed is a very small, excellent design team.

Two Top Tools: Abstraction and Modularity

About the only tools available are represented by the two truly important words in computer science: "abstraction" and "modularity." A domain, once understood, must be cut up into manageable pieces. The first step is to create levels of abstraction, hiding the details of one from another, in order to cope with the complexity. In the meantime, the function to be implemented must be divided into modules, the tricky part being to come up with the right partitioning. "How do you design the abstraction? What is the right interface?" asked Dr. Lam, remarking: "That is the hard problem."

Each new problem represents a clean slate. But what are the important concepts in the first place? How can they be put together in such a manner that they interact as little as possible and that they can be independently developed and revised? Calling this concept "the basic problem" in computer science, Dr. Lam said it must be applied at all levels. "It's not just a matter of how we come up with the component at the highest level," she explained, "but, as we take each of these components and talk about the implementation, we have to recursively apply the concepts of abstraction and modularity down to the individual functions that we write in our code."

Reusing Components to Build Systems

Through the experience of building systems, computer scientists identify components that can be reused in similar systems. Among the tasks shared by a large number of applications are, at the operating-system level, resource sharing and protections between users. Moving up the stack, protocols have been developed that allow different components to talk to each other, such as network protocols and secure communication protocols. At the next level, common code can be shared through the database, which Dr. Lam described as "a simple concept that is very, very powerful." Graphical tool kits are to be found at the next level up. Dr. Lam emphasized that difficulty decreases as one rises through the stack, and

[13] Frederick P. Brooks, *The Mythical Man-Month: Essays on Software Engineering*, 20th Anniversary Edition, New York: Addison-Wesley, 1995.

that few people are able to master the concurrency issues associated with lower-level system functions.

While the reuse of components is one attempt at solving the problem of complexity, another is the use of tools, the most important of which are probably high-level programming languages and compilers. High-level programming languages can increase software productivity by sparing programmers worry over such low-level details as managing memory or buffer overruns. Citing Java as an example of a generic programming language, Dr. Lam said the principle can be applied higher up via languages designed for more specific domains. As illustrations of such higher-level programming languages, she singled out both Matlab, for enabling engineers to talk about mathematical formulas without having to code up the low-level details, and spreadsheets, for making possible the visual manipulation of data without requiring the writing of programs. In many domains, programmers are aided in developing abstractions by using the concept of object orientation: They create objects that represent different levels of abstraction, then use these objects like a language that is tailored for the specific domain.

Individual Variations in Productivity

But even with these tools for managing complexity, many problems are left to the coders, whose job is a hard one. Dr. Lam endorsed Dr. Raduchel's assertion that productivity varies by orders of magnitude from person to person, noting that coding productivity varies widely even among candidates for the Ph.D. in computer science at Stanford. Recalling the "mythical man-month" discussed in Brooks' book, she drew an analogy to the human gestation period, asking: "If it takes one woman 9 months to produce a baby, how long would it take if you had nine women?" And Brooks' concept has darker implications for productivity, since he said that adding manpower not only would fail to make a software project finish earlier, it would in fact make a late software project even later.

But, at the same time that coding is acknowledged to be hard, "it's also deceptively simple," Dr. Lam maintained. If some of her students, as they tell her, were programming at the age of 5, how hard can programming be? At the heart of the paradox is the fact that the majority of a program has to do with handling exceptions; she drew an analogy to scuba-diving licenses, studying for which is largely taken up with "figuring out what to do in emergencies." Less than 1 percent of debugging time is required to get a program to 90 percent correctness, and it will probably involve only 10 percent of the code. That is what makes it hard to estimate software development time. Finally, in a large piece of software there is an exponential number of ways to execute the code. Experience has taught her, Dr. Lam said, that any path never before seen to have worked probably will not work. This leads to the trade-off between correctness and time to market, features, and speed. In exchange for giving up on correctness, the software developer can get the product to market more quickly, put a lot more features on it, and produce code that actually runs faster—all of which yields a cost advantage. The results of this trade-off are often seen in the consumer software market: When companies get their product to a point where they judge it to be acceptable, they opt for more features and faster time to market.

How Hardware, Software Design Handle Error

This highlights a contrast between the fields of software design and hardware design. While it may be true that hardware is orders of magnitude less complicated than software, it is still complicated enough that today's high-performance microprocessors are not without complexity issues.[14] But the issues are dealt with differently by the makers of hardware, in which any error can be very costly to the company: Just spinning a new mask, for example, may cost half a million dollars. Microprocessor manufacturers will not throw features in readily but rather introduce them very carefully. They invest heavily in CAD tools and verification tools, which themselves cost a lot of money, and they also spend a lot more time verifying or testing their code. What they are doing is really the same thing that software developers do, simulating how their product works under different conditions, but they spend more time on it. Of course, it is possible for hardware designers to use software to take up the slack and to mask any errors that are capable of being masked. In this, they have an advantage over software developers in managing complexity problems, as the latter must take care of all their problems at the software level.

"Quality assurance" in software development by and large simply means testing and inspection. "We usually have a large number of testers," Dr. Lam remarked, "and usually these are not the people whom you would trust to do the development." They fire off a lot of tests but, because they are not tied to the way the code has been written, these tests do not necessarily exercise all of the key paths through the system. While this has advantages, it also has disadvantages, one of which is that the resulting software can be "very fragile and brittle." Moreover, testers have time to fix only the high-priority errors, so that software can leave the testing process full of errors. It has been estimated that a good piece of software may contain one error every thousand lines, whereas software that is not mature may contain four and a half errors per thousand lines.

If Windows 2000, then, has 35 million lines of code, how many errors might there be? Dr. Lam recalled that when Microsoft released the program, it also—accidentally—released the information that it had 63,000 known bugs at the time of release, or about two errors per thousand lines of code. "Remember: These are

[14]There is not always a clear demarcation between hardware and software. For example, custom logic chips often contain circuitry that does for them some of the things done by software for general-purpose microprocessor chips.

the known bugs that were not fixed at the time of release," she stated. "This is not counting the bugs that were fixed before the release or the bugs that they didn't know about after the release." While it is infeasible to expect a 100 percent correct program, the question remains: Where should the line be drawn?

Setting the Bar on Bugs

"I would claim that the bar is way too low right now," declared Dr. Lam, arguing that there are "many simple yet deadly bugs" in Windows and other PC software. For an example, she turned to the problem of buffer overrun. Although it has been around for 15 years, a long time by computer-industry standards, this problem is ridiculously simple: A programmer allocates a buffer, and the code, accessing data in the buffer, goes over the bound of the buffer without checking that that is being done. Because the software does this, it is possible to supply inputs to the program that would seize control of the software so that the operator could do whatever he or she wanted with it. So, although it can be a pretty nasty error, it is a very simple error and, in comparison to some others, not that hard to fix. The problem might, in fact, have been obviated had the code been written in Java in the first place. But, while rewriting everything in Java would be an expensive proposition, there are other ways of solving the problem. "If you are just willing to spend a little bit more computational cycles to catch these situations," she said, "you can do it with the existing software," although this would slow the program down, something to which consumers might object.

According to Dr. Lam, however, the real question is not whether these errors can be stopped but who should be paying for the fix, and how much. Today, she said, consumers pay for it every single time they get a virus: The Slammer worm was estimated to cost a billion dollars, and the MS Blaster worm cost Stanford alone $800,000. The problem owes its existence in part to the monopoly status of Microsoft, which "doesn't have to worry about a competitor doing a more reliable product."

To add insult to injury, every virus to date has been known ahead of time. Patches have been released ahead of the attacks, although the lead time has been diminishing—from 6 weeks before the virus hits to, more recently, about 1 week. "It won't be long before you'll be seeing the zero-day delay," Dr. Lam predicted. "You'll find out the problem the day that the virus is going to hit you." Asking consumers to update their software is not, in her opinion, "a technically best solution."

Reliability: Software to Check the Software

A number of places have been doing research on improving software reliability, although it is of a different kind from that at the Software Engineering

Institute. From Dr. Lam's perspective, what is needed to deal with software complexity is "software to go check the software," which she called "our only hope to make a big difference." Tools are to be found in today's research labs that, running on existing codes such as Linux, can find thousands of critical errors—that is, errors that can cause a system to crash. In fact, the quality of software is so poor that it is not that hard to come up with such tools, and more complex tools that can locate more complex errors, such as memory leaks, are also being devised. Under study are ideas about detecting anomalies while the program runs so that a problem can be intercepted before it compromises the security of the system, as well as ideas for higher-level debugging tools. In the end, the goal should not be to build perfect software but software that can automatically recover from some errors. Drawing a distinction between software tools and hardware tools, she asserted that companies have very little economic incentive to encourage the growth of the former. If, however, software producers become more concerned about the reliability of their product, a little more growth in the area of software tools may ensue.

In conclusion, Dr. Lam reiterated that many problems arise from the fact that software engineering, while complex and hard, is at the same time deceptively simple. She stressed her concern over the fact that, under the reigning economic model, the cost of unreliability is passed to the unwitting consumer and there is a lack of investment in developing software tools to improve the productivity of programmers.

INTRODUCTION

James Socas
Senate Committee on Banking

Introducing the next speaker, Dr. Hal Varian, Mr. Socas noted that Dr. Varian is the Class of 1944 Professor at the School of Information Management and Systems of the Haas School of Business at the University of California at Berkeley, California; co-author of a best-selling book on business strategy, *Information Rules: A Strategic Guide to the Network Economy*; and contributor of a monthly column to the *New York Times*.

OPEN-SOURCE SOFTWARE

Hal R. Varian
University of California at Berkeley

While his talk would be based on work he had done with Carl Shapiro focusing specifically on the adoption of Linux in the public sector, Dr. Varian noted that much of that work also applies to Linux or open-source adoption in general.[15] Literature on the supply of open source addresses such questions as who creates it, why they do it, what the economic motivations are, and how it is done.[16] A small industry exists in studying CVS logs: how many people contribute, what countries they're from, a great variety of data about open-source development.[17] The work of Josh Lerner of Harvard Business School, Jean Tirole, Neil Gandal, and several others provides a start in investigating that literature.[18] Literature on the demand for open source addresses such questions as who uses it, why they use it, and how they use it. While the area is a little less developed, there are some nice data sources, including the FLOSS Survey, or Free/Libre/Open-Source Software Survey, conducted in Europe.[19]

Varian and Shapiro's particular interest was looking at economic and strategic issues involving the adoption and use of open-source software with some focus on the public sector. Their research was sponsored by IBM Corporation, which, as Dr. Varian pointed out, has its own views on some of the issues studied. "That's supposed to be a two-edged sword," he acknowledged. "One edge is that that's who paid us to do the work, and the other edge is that they may not agree with what we found."

Definition of Open Source

Distinguishing open-source from commercial software, Dr. Varian defined open source as software whose source code is freely available. Distinguishing open interface from proprietary interface, he defined an open interface as one that is completely documented and freely usable, saying that could include the pro-

---

[15]See Hal R. Varian and Carl Shapiro, "Linux Adoption in the Public Sector: An Economic Analysis," Department of Economics, University of California at Berkeley, December 1, 2003. Accessed at <*http://www.sims.berkeley.edu/~hal/Papers/2004/linux-adoption-in-the-public-sector.pdf*>.

[16]An extended bibliography on open-source software has been compiled by Brenda Chawner, School of Information Management, Victoria University of Wellington, New Zealand. Accessed at <*http://www.vuw.ac.nz/staff/brenda_chawner/biblio.html*>.

[17]CVS (Concurrent Versions System) is a utility used to keep several versions of a set of files and to allow several developers to work on a project together. It allows developers to see who is editing files, what changes they made to them, and when and why that happened.

[18]See Josh Lerner and Jean Tirole, "The Simple Economics of Open Source," Harvard Business School, February 25, 2000. Accessed at <*http://www.hbs.edu/research/facpubs/*>. See also Neil Gandal and Chaim Fershtman, "The Determinants of Output per Contributor in Open Source Projects: An Empirical Examination," CEPR Working Paper 2650, 2004. Accessed at <*http://spirit.tau.ac.il/public/gandal/Research.htmworkingpapers/papers2/9900/00-059.pdf*>.

[19]The FLOSS surveys were designed to collect data on the importance of open source software (OSS) in Europe and to assess the importance of OSS for policy- and decision-making. See FLOSS Final Report, accepted by the European Commission in 2002. Accessed at <*http://www.infonomics.nl/FLOSS/report/index.htm*>.

grammer interface, the so-called Application Programming Interface or API; the user interface; and the document interface. Without exception, in his judgment, open-source software has open interfaces; proprietary software may or may not have open interfaces.[20] Among the themes of his research is that much of the benefit to be obtained from open-source software comes from the open interface, although a number of strategic issues surrounding the open interface mandate caution. Looking ahead to his conclusion, he called open-source software a very strong way to commit to an open interface while noting that an open interface can also be obtained through other sorts of commitment devices. Among the many motivations for writing software—"scratching an itch, demonstrating skill, generosity, [and] throwing it over the fence" being some he named—is making money, whether through consulting, furnishing support-related services, creating distributions, or providing complements.

The question frequently arises of how an economic business can be built around open source. The answer is in complexity management at the level of abstraction above that of software. Complexity in production processes in organizations has always needed to be managed, and while software is a tool for managing complexity, it creates a great deal of complexity itself. In many cases—in the restaurant business, for instance—there is money to be made in combining standard, defined, explicit ingredients and selling them. The motor vehicle provides another example: "I could go out and buy all those parts and put them together in my garage and make an automobile," Dr. Varian said, "but that would not be a very economic thing to do." In software as well, there is money to be made by taking freely available components and combining them in ways that increase ease of management, and several companies are engaged in doing that.

The Problem of Forking or Splintering

The biggest danger in open-source software is the problem of forking or splintering, "a la UNIX." A somewhat anarchic system for developing software may yield many different branches of that software, and the challenge in many open-source software projects is remaining focused on a standard, interchangeable, interoperable distribution. Similarly, the challenge that the entire open-source industry will face in the future is managing the forking and splintering problem.

As examples of open source Dr. Varian named Linux and BSD, or Berkeley Standard Distribution. Although no one knows for certain, Linux may have

[20]A key question for software designers is where to put an interface? That is, what is the modularity? If open-source software made its interfaces open, but chose to put in fewer or no internal interfaces, that could be significant. Another issue concerns the stability of the interface over time. If, for example, open-source interfaces are changed more readily, would this create more challenges for complementors?

18 million users, around 4 percent of the desktop market. Many large companies have chosen Linux or BSD for their servers; Amazon and Google using the former, Yahoo the latter. These are critical parts of their business because open source allows them to customize the application to their particular needs. Another prominent open-source project, Apache, has found through a thorough study that its Web server is used in about 60 percent of Web sites. "So open source is a big deal economically speaking," Dr. Varian commented.

What Factors Influence Open-source Adoption?

Total cost of ownership. Not only the software code figures in, but also support, maintenance, and system repair. In many projects the actual expense of purchasing the software product is quite low compared to the cost of the labor necessary to support it. Several studies have found a 10 to 15 percent difference in total cost of ownership between open-source and proprietary software—although, as Dr. Varian remarks, "the direction of that difference depends on who does the study" as well as on such factors as the time period chosen and the actual costs recorded. It is also possible that, in different environments, the costs of purchasing the software product and paying the system administrators to operate it will vary significantly. Reports from India indicate that a system administrator is about one-tenth of the cost of a system administrator in the U.S., a fact that could certainly change the economics of adoption; if there is a 10 to 15 percent difference in total cost of ownership using U.S. prices, there could be a dramatic difference using local prices when labor costs are taken into account.

Switching costs. Varian and Shapiro found this factor, which refers to the cost incurred in switching to an alternative system with similar functionality, to be the most important. Switching tends to cost far more with proprietary interfaces for the obvious reason that it requires pretty much starting from scratch. When a company that has bought into a system considers upgrading or changing, by far the largest cost component it faces is involved in retraining, changing document formats, and moving over to different systems. In fact, such concerns dominate that decision in many cases, and they are typically much lower for software packages with open interfaces.

Furthermore, cautioned Dr. Varian, "vendors—no matter what they say—will typically have an incentive to try to exploit those switching costs." They may give away version n of a product but then charge what the market will bear for version $n + 1$. While this looms as a problem for customers, it also creates a problem for vendors. For the latter, one challenge is to convince the customer that, down the road, they will not exploit the position they enjoy as not only supplier of the product but also the sole source of changes, upgrades, updates, interoperability, and so on. One way of doing so is to have true open interfaces, and since open source is a very strong way to achieve the open interface that customers demand, it is attractive as a business model.

By way of illustration, Dr. Varian invoked the history of PostScript, whose origin was as a page-description language called Interleaf that was developed by Xerox. The company wanted to release the system as a proprietary product, but potential adopters, afraid of locking themselves into Xerox, shied away from buying it. Interleaf's developer left Xerox to start Adobe; developed a similar language, PostScript; and released its specification into the public domain, which allowed anybody to develop a PostScript interpreter. Creating a competitive environment was necessary because, unless they had a fallback, customers would not commit to the product.

The Limits of Monopoly Power

According to Dr. Varian, this has increasingly become an issue in software systems, to the point that in many cases it is hard to exploit a monopoly position. Microsoft, for instance, faces an extremely difficult business-strategy problem when it comes to the Chinese market. "Why should [China's government] allow users there to take the first shot of a very expensive drug?" he asked. "It's cheap now—maybe it's free because it's a pirated copy—but in the future it will be expensive." Since most communication in China is still domestic and from Chinese to Chinese, the network effects are relatively small. So the government has an incentive to choose the system with the lowest switching cost, then build its own network for document interchange. Microsoft, for its part, faces a dilemma; perhaps it can tolerate, even encourage piracy for a while to try to get the system adopted widely on the ground. But how can it commit to refraining from future exploitation?

Mandating open interface, now being discussed by the European Union, would be "a big help" in solving this problem. Because mandating open interface is also a strong way of committing to lowering switching costs, it is in many ways attractive from the point of the adopting company, although it could spill over into existing markets and cut into profits in incumbent markets.

Reasons for Adopting Open Source

The FLOSS Survey previously mentioned, having investigated reasons for adoption of open source, argued that higher stability was very important and that maintainability was also important because the structure of the software was more attractive. Since the software is developed in a very distributed manner, it has to be more modular because of the nature of the production process.[21] Modular

[21]Most of the modules developed by non-core Linux programmers are device drivers and other plug-ins.

software is typically more robust, and forcing modularity through the development process may simplify maintenance.

Usability is another interesting case. User testing is typically very labor intensive and costly, but the intellectual property status of interfaces is still somewhat ambiguous. Dr. Varian recalled that the U.S. Supreme Court split 4-4 on the Lotus-Quattro case, and the question of to just what degree a user interface may be copied is not clear under U.S. laws.[22] In addition, it may be possible for investment in user testing to be appropriated by other vendors. Indeed, many open-source projects look very similar, with respect to the user interface, to some well-known proprietary products.

Then there's the issue of security. One advantage of having access to the source code is that it permits customizing product to meet specific goals. On its Web site, NSA posts a hardened version of Linux that can be downloaded and used for certain applications, and that can be made smaller, larger, harder, more secure, more localized, and so on. Such flexibility is an advantage.

Turning to licensing, Dr. Varian noted that there are many different licenses and referred to a paper in a legal journal that distinguished some 45 of them. Perhaps the most notorious is the GNU public license, which has a provision, "Copyleft," requiring those who modify and distribute open-source software outside their organization to include the source code. While this is a somewhat controversial provision, it applies only under narrow conditions and is in some respect not so different from any other kind of intellectual property right: It is simply a provision of the license that the creator has put into the product.

Open Source and Economic Development

Because this original focus of Varian and Shapiro's work was on adoption in the public sector, it treats claims about the importance of open-source software for economic development. The first and most prominent piece of advice that one might give, Dr. Varian stated, is to favor open interfaces wherever possible. He reiterated that while the open interface is provided by open source, there are other ways to get software with open interfaces, and much proprietary software has relatively open interfaces. He also recommended being very careful about the lock-in problem: "If you're going to have switching costs down the road from adopting, let's say, something with a proprietary interface on any of the sides I mentioned, then it's going to be difficult in terms of the interoperability." Recalling Dr. Raduchel's praise of standards, he insisted on a single qualification: Such

[22]The Supreme Court in 1996 was unable to decide the highly publicized case of Lotus vs. Borland, which turned over whether the menu and command structure of the Lotus 1-2-3 spreadsheet program could be copyrighted. Lotus sued Borland in 1990, seeking $100 million in damages because Borland's Quattro spreadsheet mimicked the 1-2-3 commands.

examples as the flathead screw and the resolution of U.S. TV show that premature standardization can lock users into a standard that later they wish they hadn't chosen. Progress stops in a component when it becomes standardized and everybody produces to that standard. This can spur innovation, because this component can be used to build other, more complex systems. But there is always a choice—which in many cases is very agonizing—as to whether that is the right point at which to standardize the particular technology.

Another constructive role played by open-source software, said Dr. Varian, is in education: "Imagine trying to train mechanics if they could never work with real engines!" This point was championed by Donald Knuth through his release of TeX, a large open-source program. Knuth was concerned that students could never get access to real programs; they were unable to see how proprietary programs worked, and the programs they could see, written by physicists, provided terrible instruction for computer scientists. Knuth's aim was to write a program that, because it was good from the viewpoint of design and structure, could serve as a model for future students. A great deal of open-source software has the side effect of being extremely valuable in training and education. According to the Berkeley computer science department's chair, all of its students use Linux and many claim not to know Windows at all. Admittedly, this may be accounted for partly by pride and partly by a desire to avoid being dragooned into fixing their friends' systems. But what is important is that Linux is the model they use for software development, something that also applies in less-developed countries.

In closing, Dr. Varian pointed to a paper posted on his Web site that is accompanied by discussion of economic effects of open-source software, including the issues of complementarities, commitment, network effects, licensing terms, and bundling (See Figure 4).[23] He then invited questions.

DISCUSSION

Dr. Varian was asked whether the fact that today's scientists typically do not understand the software they use implies limits on their understanding of the science in which they are engaged.

Who Should Write the Software?

Acknowledging this as a perennial problem, Dr. Varian pointed to a related phenomenon that can be observed at most universities: While there is a separate statistics department, individual departments have internal statistics departments that pursue, for example, psychometrics, econometrics, biometrics, or public-

[23]Professor Varian's papers can be accessed at <*http://www.sims.berkeley.edu/~hal/people/hal/papers.html*>.

EFFECT	DEFINITION	IMPLICATION
Complementarity	The value of an operating system depends on availability of applications.	Consider the entire system of needs before making choice.
Switching costs	The cost of switching any one component of an IT system can be very high.	Make choices that preserve your flexibility in the future.
Commitment	Vendors may promise flexibility or low prices in the future but not deliver.	Look for firm commitments from vendors, such as a commitment to open interfaces.
Network effects	The value of an application or operating system may depend heavily on how many other users adopt it.	For a closed network of users, standardization within the network is more important than choosing an industry standard.
Licensing terms	A perpetual license involves a one-time payment; a subscription involves a yearly payment.	Licenses can be particularly pernicious when switching costs are high.
Bundling	Vendors will want to sell software in bundles to make future entry into the market difficult.	Purchasing a bundle now may reduce your future costs, but will also limit your flexibility and choices.

FIGURE 4 Economic effects.

health statistics. All sciences need to know something about software engineering in order to build systems that can meet their needs, but is it better that they build them themselves, outsource them to somebody else, or strike a more collaborative arrangement? Berkeley is trying an interdisciplinary approach.

Dr. Lam, offering a computer scientist's perspective, said that computer science people don't necessarily know any better than scientists in other disciplines what their software is doing. "Otherwise," she asked, "why do we have this software problem?" Whether the software is open source or not is not the issue. The fact is that, after writing a program, one can look at all the lines of code and still not know what the software is doing. Even the more readily accessible high levels will behave differently than expected because there are going to be bugs in them. Part of the object of current research is achieving a fuller view of what the code is actually doing so that one is not just staring at the source lines.

Work Organization for Software Development

Mr. Socas, alluding to his experience working in Silicon Valley, recalled that one of the big debates was on how to organize people for software development.

In competition were the industrial or factory model, which is the way most of the U.S. economy is set up, and the "Hollywood" or campus model, which assumes that the skill set needed to create software is unique and that, therefore, those with the talent to do it should be given whatever it is they want. He asked the panelists to comment on the organizational challenges remaining to optimal software development and on whether open-source software changes the game by providing an alternative to the hierarchical, all-in-one company model.

Dr. Varian, disclaiming expertise in the matter, nonetheless remarked that the success of the distributed model of open-source development—with contributions from around the world being merged into a product that actually works—has come as a big surprise to some. He mentioned the existence of a Pareto distribution or power law for the field under which a relatively small number of people contribute a "big, big chunk of the code," after which bug fixes, patches, and other, minor pieces are contributed "here or there."[24] This underlines the importance of the power programmer: the principle that a few very good people, made more accessible by the advent of the Internet, can be far more productive than many mediocre people. Returning to the question of China, he posited that if that nation "really does go open source in a very big way," the availability of hundreds of millions of Chinese engineers for cleaning up code could have a major and beneficial impact on the rest of the world.

Open-Source vs. Proprietary in the Future

Asked to speculate on the relative dominance of open-source and proprietary platforms a decade down the road, Dr. Varian predicted that open source would make a strong impact. This will be particularly true in parts of the world where networks are not yet in place, as building to an open standard is easier and far better for long-run development. With China, India, and other countries making major efforts in this area, the United States and, to a lesser extent, Europe will be on their own path, using a different standard from the rest of the world. He acknowledged this as "an extreme prediction," but added: "That would be my call at this point."

Kenneth Walker of SonicWALL noted that, although it is not widely known, the Mac OS (operating system) was built on top of BSD.[25] Open source, there-

---

[24]See A. Mockus, R. Fielding, and J. D. Herbsleb, "Two Case Studies of Open Source Software Development: Apache and Mozilla," *ACM Transactions on Software Engineering and Methodology,* 11(3):309-346, 2002.

[25]In the middle 1970s, AT&T began to license its Unix operating system. At little or no cost, individuals and organizations could obtain the C source code. When the University of California at Berkeley received the source code, Unix co-creator Ken Thompson was there as visiting faculty. With his help, researchers and students, notably Sun co-founder Bill Joy, improved the code and developed the Berkeley Software Distribution (BSD). Funded by a grant from DARPA, the Berkeley Computer

fore, is "the kernel of what the Mac is," and even the changes that Apple has made to that kernel and core have been put out by the company as an open-source project called Darwin—which, "interestingly enough," runs on Intel hardware.

Rights Specified in Copyright Licenses

Dr. Varian was asked for his recommendations on what should not be allowed in a license that constrains subsequent use of software should Congress move to revise copyright laws to state what types of licenses copyright owners of software should and should not be able to specify.

While declining to make strong recommendations for the time being, Dr. Varian pointed out that software is most valuable when it can be combined, recombined, and built upon to produce a secure base upon which additional applications can in turn be built. The challenge is ensuring the existence of incentives sufficient to developing those basic components, while at the same time ensuring that, once they are built, they will be widely available at low cost so that future development is stimulated. Of major concern are licenses that have some ex-post manipulability, he said, describing what economists call "holdup": "I release a license, and you use it for 2 years. When it's up for renewal I say, 'Wow, you've built a really fantastic product out of this component—so I'm going to triple the price.' " Since holdup discourages widespread use of a product, it is important that licenses, whatever form they take, can be adopted with some confidence that the path to future use of the product is secure.

Who Uses Open-Source, and Where Are They?

Asked by Egils Milbergs of the Center for Accelerating Innovation to characterize the users of open-source software, Dr. Varian said that very few are end users and that many products developed in open source are meant for developers. He explained this by elaborating on one of the motivations he had listed for writing software, "scratching an itch:" A software professional confronted by a particular problem builds a tool that will help solve it, then decides to make the tool available to others to use and improve. While there have been many efforts to make software more user-friendly down to the end-user level, that is often a very difficult task. Since copying existing interfaces may be the easiest way to accomplish this, it is what has commonly been done.

In response to another question from Mr. Milbergs, on the geographic distribution of open-source users, Dr. Varian noted that logs from CVS, a system used

---

Systems Research Group (CSRG) was the most important source of Unix development outside of Bell Labs. Along with AT&T's own System V, BSD became one of the two major Unix flavors. See <*http://kb.iu.edu/data/agom.html*>.

to manage the development of open-source software, are an important source of data. These publicly available logs show how many people have checked in, their email addresses, and what lines they have contributed. Studies of the logs that examine geographic distribution, time distribution, and lines contributed paint a useful picture of the software-development process.

Assigning Value to Freely Distributed Code

David Wasshausen of the U.S. Department of Commerce asked how prevalent the use of open-source code currently is in the business world and how value is assigned both to code that is freely distributed and to the final product that incorporates it. As an economic accountant, he said, he felt that there should be some value assigned to it, adding, "It's my understanding that just because it's open-source code it doesn't necessarily mean that it's free."[26] He speculated that the economic transaction may come in with the selling of complete solutions based on open-source code, citing the business model of companies like Red Hat.

Dr. Varian responded by suggesting that pricing of open-source components be based on the "next-best alternative." He recounted a recent conversation with a cash-register vendor who told him that the year's big innovation was switching from Windows 95 to Windows 98—a contemporary cash register being a PC with an alternative front end. Behind this switch was a desire for a stable platform which had known bugs and upon which new applications could be built. Asked about Linux, the vendor replied that the Europeans were moving into it in a big way. Drawing on this example, Dr. Varian posited that if running the cash register using Linux rather than Windows saved $50 in licensing fees, then $50 would be the right accounting number. He acknowledged that the incremental cost of developing and supporting the use of Linux might have to be added. However, he likened the embedding of Linux in a single-application device such as a cash register to the use of an integrated circuit or, in fact, of any other standardized part. "Having a component that you can drop into your device, and be pretty confident it's going to work in a known way, and modify any little pieces you need to modify for that device," he said, "is a pretty powerful thing." This makes using Linux or other open-source software extremely attractive to the manufacturer who is building a complex piece of hardware with a relatively simple interface and needs no more than complexity management of the electronics. In such a case, the open-source component can be priced at the next-best alternative.

---

[26]The presence of open-source software implicitly raises a conundrum for national economic accountants. If someone augments open-source code without charge, it will not have a price, and thus will not be counted as investment in measures of national product. This reflects the national economic accounting presumption of avoiding imputations, especially when no comparable market transactions are available as guides.

INTRODUCTION

James Socas
Senate Committee on Banking

Introducing Kenneth Walker of SonicWALL, Mr. Socas speculated that his title—Director of Platform Evangelism—is one that "you'll only find in Silicon Valley."

MAKING SOFTWARE SECURE AND RELIABLE

Kenneth Walker
SonicWALL

Alluding to the day's previous discussion of the software stack, Mr. Walker stated as the goal of his presentation increasing his listeners' understanding of the security stack and of what is necessary to arrive at systems that can be considered secure and reliable. To begin, he posed a number of questions about what it means to secure software:

- What is it we're really securing?
- Are we securing the individual applications that we run: Word? our Web server? whatever the particular machinery-control system is that we're using?
- Or are we securing access to the machine that that's on?
- Or are we securing what you do with the application? the code? the data? the operating system?

All of the above are involved with the security even of a laptop, let alone a network or collection of machines and systems that have to talk to each other—to move data back and forth—in, hopefully, a secure and reliable way.

Defining Reliability in the Security Context

The next question is how to define reliability. For some of SonicWALL's customers, reliability is defined by the mere fact of availability. That they connect to the network at all means that their system is up and running and that they can do something; for them, that amounts to having a secure environment. Or is reliability defined by ease of use? by whether or not the user has to reboot all the time? by whether there is a backup mechanism for the systems that are in place? Mr. Walker restated the questions he raised regarding security, applying them to reliability: "Are we talking about applications? or the operating system? or the machine? or the data that is positively, absolutely needed right now?"

Taking up the complexity of software, Mr. Walker showed a chart depicting the increase in the number of lines of code from Windows 3.1 through Windows

XP (See Figure 5). In the same period, attacks against that code—in the form of both network-intrusion attempts and infection attempts—have been growing "at an astronomical pace" (See Figure 6). When Code Red hit the market in 2001, 250,000 Web servers went down in 9 hours, and all corners of the globe were affected. The Slammer worm of 2003 hit faster and took out even more systems.

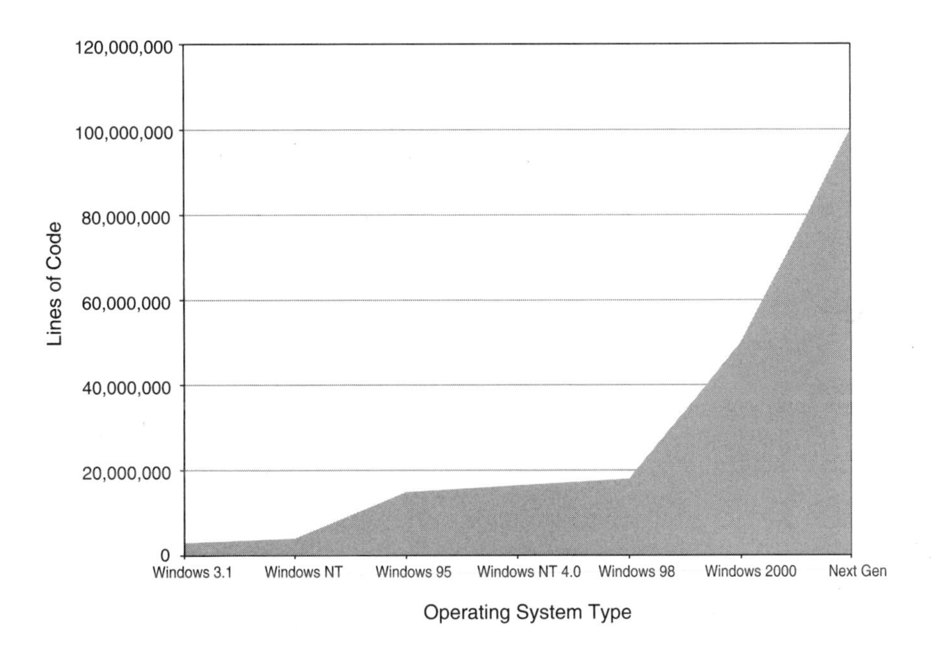

FIGURE 5 Software is getting more complex.

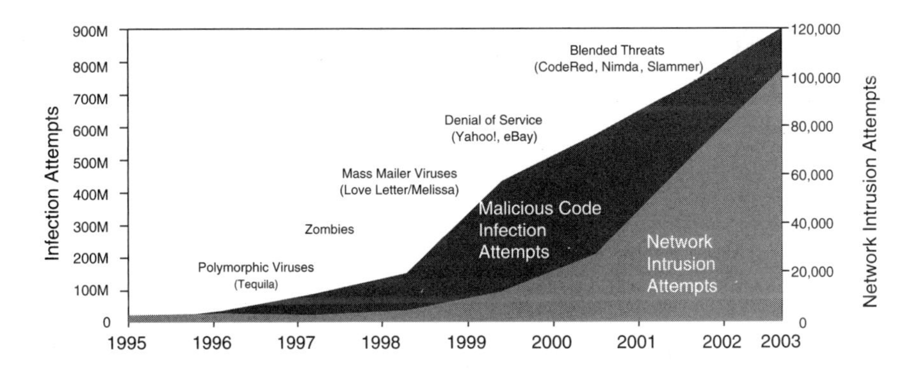

FIGURE 6 Attacks against code are growing.
NOTE: Analysis by Symantec Security Response using data from Symantec, IDC & ICSA.

Threats to Security from Outside

These phenomena are signs of a change in the computing world that has brought with it a number of security problems Mr. Walker classified as "external," in that they emanate from the environment at large rather than from within the systems of individual businesses or other users. In the new environment, the speed and sophistication of cyber-attacks are increasing. The code that has gone out through the open-source movement has not only helped improve the operating-system environment, it has trained the hackers, who have become more capable and better organized. New sorts of blended threats and hybrid attacks have emerged.[27]

Promising to go into more detail later on the recent Mydoom attack, Mr. Walker noted in passing that the worm did more than infect individuals' computers, producing acute but short-lived inconvenience, and then try to attack SCO, a well-known UNIX vendor.[28] It produced a residual effect, to which few paid attention, by going back into each machine's settings; opening up ports on the machine; and, if the machine was running different specific operating systems, making it more available to someone who might come in later, use that as an exploit, and take over the machine for other purposes—whether to get at data, reload other code, or do something else. The attack on SCO, he cautioned, "could be considered the red herring for the deeper issue." Users could have scrubbed their machines and gotten Mydoom out of the way, but the holes that the worm left in their armor would still exist.

Returning to the general features of the new environment, Mr. Walker pointed out that the country has developed a critical infrastructure that is based on a network. Arguing that the only way to stop some security threats would be to kill the Internet, he asked how many in the audience would be prepared to give up email or see it limited to communication between departments, and to go back to faxes and FedEx. Concluding his summary of external security issues, he said it was important to keep in mind that cyber-attacks cross borders and that they do not originate in any one place.

---

[27]The term "hacker," is defined by the Webopedia as "A slang term for a computer enthusiast, i.e., a person who enjoys learning programming languages and computer systems and can often be considered an expert on the subject(s)." Among professional programmers, depending on how it used, the term can be either complimentary or derogatory, although it is developing an increasingly derogatory connotation. The pejorative sense of hacker is becoming more prominent largely because the popular press has co-opted the term to refer to individuals who gain unauthorized access to computer systems for the purpose of stealing and corrupting data. Many hackers, themselves, maintain that the proper term for such individuals is "cracker."

[28]Maureen O'Gara, "Huge MyDoom Zombie Army Wipes out SCO," *Linux World Magazine*, February 1, 2004.

Threats to Security from Inside

Internal security issues abound as well. Although it is not very well understood by company boards or management, many systems now in place are misconfigured, or they have not had the latest patches applied and so are out of date. Anyone who is mobile and wireless is beset by security holes, and mobile workers are extending the perimeter of the office environment. "No matter how good network security is at the office, when an employee takes a laptop home and plugs up to the network there, the next thing you know, they're infected," Mr. Walker said. "They come back to the office, plug up, and you've been hacked." The outsourcing of software development, which is fueled by code's growing complexity, can be another source of problems. Clients cannot always be sure what the contractors who develop their applications have done with the code before giving it back. It doesn't matter whether the work is done at home or abroad; they don't necessarily have the ability to go through and test the code to make sure that their contractor has not put back doors into it that may compromise their systems later.

Mr. Walker stressed that there is no one hacker base—attacks have launched from all over the globe—and that a computer virus attack can spread instantaneously from continent to continent. This contrasts to a human virus outbreak, which depends on the physical movement of individuals and is therefore easier to track.

Taking Responsibility for Security

While there is a definite problem of responsibility related to security, it is a complicated issue, and most people do not think about it or have not thought about it to date. Most reflexively depend on "someone in IT" to come by with a disk to fix whatever difficulties they have with their machines, never pausing to think about how many other machines the information specialist may have to visit. Similarly, managers simply see the IT department as taking care of such problems and therefore rarely track them. But "prevention, unfortunately, takes people to do work," said Mr. Walker, adding, "We haven't gotten smart enough around our intrusion-prevention systems to have them be fully automated and not make a lot of mistakes." The frequency of false positives and false negatives may lead businesses to throw away important protections against simple threats.

Perceptions of Security

Masking the amount of work that goes into security is the fact that the user can tell security is working correctly only when nothing goes wrong, which makes adequate security appear easy to attain. Masking its value is that no enterprise gets any economic value for doing a good job securing itself; the economic value

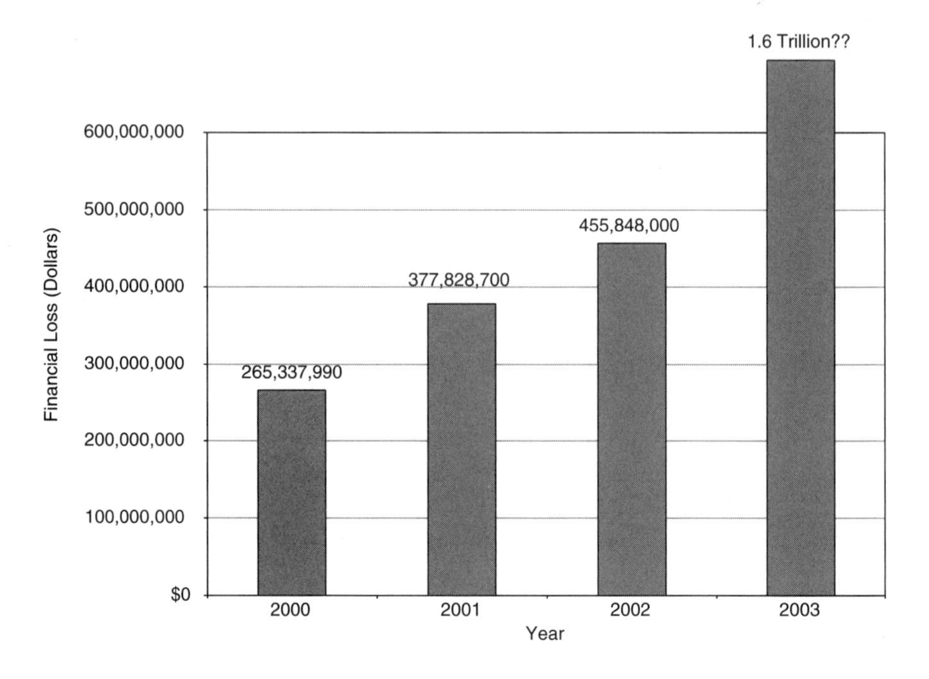

FIGURE 7 Financial impact of worms and viruses.

becomes apparent only when a business starts to be completely taken down by not having invested in security.[29] For this reason, time and money spent on security are often perceived as wasted by managers, whose complaint Mr. Walker characterized as " 'I've got to put in equipment, I've got to train people in security, I've got to have people watching my network all the time.' " The St. Paul Group, in fact, identified the executive's top concern in 2003 as "How do I cut IT spending?" Unless more money is spent, however, enterprise security will remain in its current woeful state, and the financial impact is likely to be considerable. *E-Commerce Times* has projected the global economic impact of 2003's worms and viruses at over $1 trillion (See Figure 7).

[29]As noted by a reviewer, Mr. Walker's discussion of threats to security, including hacking, are comparable to national economic accounts' failure to distinguish purely defensive expenditures, such as the hiring of bank guards to prevent robberies. There has been some discussion of measuring such things as robberies as "bads," which are negative entries that reduce value of banking services, but the international economic accounting guidelines have not yet successfully addressed the issue. Until there is more general agreement as to how to account for bads such as robberies and pollution, it is unlikely that national economic accounts will be able to deal with such issues.

The Mydoom Virus and its Impact

The first attack of 2004, Mydoom, was a mass-mailing attack and a peer-to-peer attack. Mr. Walker described it as "a little insidious" in that, besides hitting as a traditional email worm, it tried to spread by copying itself to the shared directory for Kazaa clients when it could find it. Infected machines then started sharing data back and forth even if their email was protected. "The worm was smart enough to pay attention to the social aspects of how people interact and share files, and wrote itself into that," he commented. Mydoom also varied itself, another new development. Whereas the Code Red and Slammer worms called themselves the same thing in every attachment, Mydoom picked a name and a random way of disguising itself: Sometimes it was a zip file, sometimes it wasn't. It also took the human element into account. "As much as we want to train our users not to," Mr. Walker observed, "if there's an attachment to it, they double-click." There was a woman in his own company who clicked on the "I Love You" variant of Slammer 22 times "because this person was telling her he loved her, and she really wanted to see what was going on, despite the fact that people were saying, 'Don't click on that.' "

As for impact, Network Associates estimated Mydoom and its variants infected between 300,000 and 500,000 computers, 10 to 20 times more than the top virus of 2003, SoBig. F-Secure's estimate was that on January 28, 2004, Mydoom accounted for 20 to 30 percent of global email traffic, well above previous infections. And, as mentioned previously, there were after-effects: MyDoom left ports open behind it that could be used as doorways later on. On January 27, 2004, upon receipt of the first indications of a new worm's presence, Network Associates at McAfee, Symantec, and other companies diligently began trying to deconstruct it, figure out how it worked, figure out a way to pattern-match against it, block it, and then update their virus engines in order to stop it in its tracks. Network Associates came out with an update at 8:55 p.m., but, owing to the lack of discipline when it comes to protecting machines, there was a peak of infection the next morning at 8:00, when users launched their email and started to check the messages (See Figure 8). There are vendors selling automated methods of enforcing protection, but there have not been many adopters. "People need to think about whether or not enforced antivirus in their systems is important enough to pay for," Mr. Walker stated.

The Hidden Costs of Security Self-Management

Self-management of antivirus solutions has hidden costs. For Mydoom, Gartner placed the combined cost of tech support and productivity lost owing to workers' inability to use their machines at between $500 and $1,000 per machine infected. Mr. Walker displayed a lengthy list of cost factors that must be taken into account in making damage estimates and commented on some of these elements:

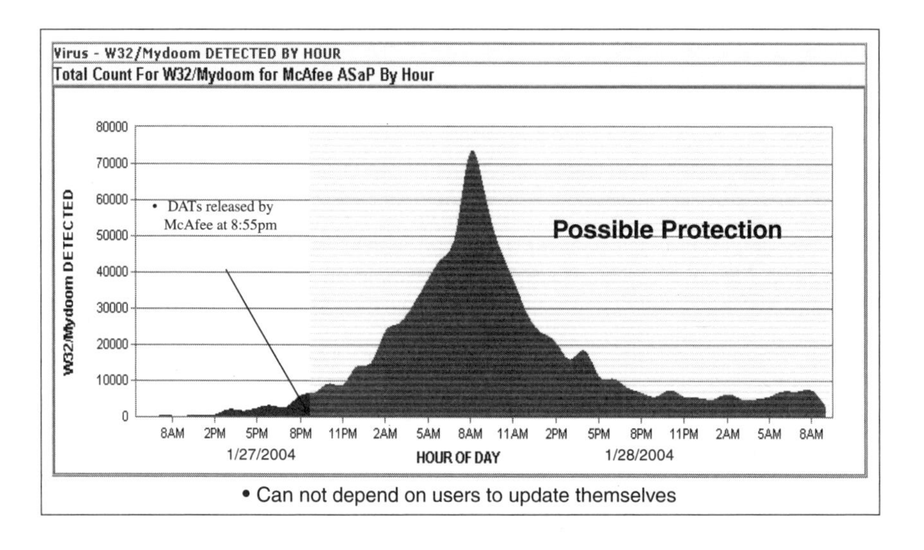

FIGURE 8 Antivirus enforcement in action.

- **help-desk support** for those who are unsure whether their machines have been affected, to which the expense of 1-800 calls may be added in the case of far-flung enterprises;
- **false positives**, in which time and effort are expended in ascertaining that a machine has not, in fact, been infected;
- **overtime payments** to IT staff involved in fixing the problem;
- **contingency outsourcing** undertaken in order to keep a business going while its system is down, an example being SCO's establishing a secondary Web site to function while its primary Web site was under attack;
- **loss of business**;
- **bandwidth clogging**;
- **productivity erosion**;
- **management time reallocation**;
- **cost of recovery**; and
- **software upgrades**.

According to mi2g consulting, by February 1, 2004, Mydoom's global impact had reached $38.5 billion.

Systems Threats: Forms and Origins

The forms of system threats vary with their origins. The *network attack* targets an enterprise's infrastructure, depleting bandwidth and degrading or compromising online services. Such an attack is based on "the fact that we're sharing a network experience," Mr. Walker commented, adding, "I don't think any of us really wants to give that up." In an *intrusion*, rather than bombarding a network the attacker tries to slide into it surreptitiously in order to take control of the system—to steal proprietary information, perhaps, or to alter or delete data. Mr. Walker invited the audience to imagine a scenario in which an adversary hired a hacker to change data that a company had attested as required under the Sarbanes-Oxley Act, then blew the whistle on the victim for non-compliance. Defining *malicious code*, another source of attack, as "any code that engages in an unwanted and unexpected result," he urged the audience to keep in mind that "software security is not necessarily the same thing as security software."

Attacks also originate in what Mr. Walker called *people issues*. One variety, the *social engineering attack*, depends on getting workers to click when they shouldn't by manipulating their curiosity or, as in the case of the woman who opened the "I Love You" virus 22 times, more complex emotions. This form of attack may not be readily amenable to technical solution, he indicated. *Internal resource abuses* are instances of employees' using resources incorrectly, which can lead to a security threat. In *backdoor engineering*, an employee "builds a way in [to the system] to do something that they shouldn't do."

Proposing what he called "the best answer overall" to the problem of threats and attacks, Mr. Walker advocated layered security, which he described as a security stack that is analogous to the stack of systems. This security stack would operate from the network gateway, to the servers, to the applications that run on those servers. Thus, the mail server and database servers would be secured at their level; the devices, whether Windows or Linux or a PDA or a phone, would be secured at their level; and the content would be secured at its level (See Figure 9). Relying on any one element of security, he argued would be "the same as saying, 'I've got a castle, I've got a moat, no one can get in.' But when somebody comes up with a better catapult and finds a way to get around or to smash my walls, I've got nothing to defend myself." The security problem, therefore, extends beyond an individual element of software.

DISCUSSION

Charles Wessner of the National Research Council asked Mr. Walker whether he believed in capital punishment for hackers and which risks could be organized governmentally.

Punishment or Reward for Hackers?

While doubting that hackers would be sentenced to death in the United States, Mr. Walker speculated that governments elsewhere might justify "extremely brutal" means of keeping hackers from their territory by citing the destruction

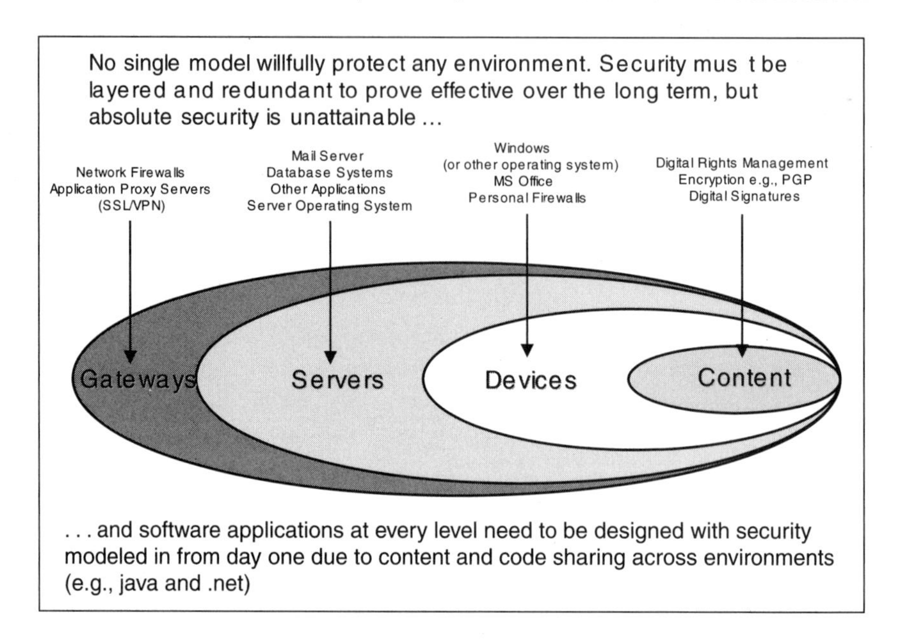

FIGURE 9 The solution: layered security.

that cyber-attacks can wreak on a national economy. He pointed to Singapore's use of the death penalty for drug offenses as an analogue.

Following up, Dr. Wessner asked whether the more usual "punishment" for hackers wasn't their being hired by the companies they had victimized under a negotiated agreement.

Mr. Walker assented, saying that had been true in many cases. Hackers, he noted, are not always motivated by malice; sometimes they undertake their activities to test or to demonstrate their skill. He called it unfortunate that, in the aftermath of 9/11, would-be good Samaritans may be charged with hacking if, having stumbled into a potential exploit, they inform the company in question of the weakness they have discovered.

Dr. Lam protested against imposing severe punishment on the hackers. "What happens is that somebody detects a flaw, posts that information—in fact, tells the software companies that there is such an error and gives them plenty of time to release a patch—and then this random kid out there copies the code that was distributed and puts it into an exploit." Rejecting that the blame lies with the kid, she questioned why a company would distribute a product that is so easily tampered with in the first place.

Concurring, Mr. Walker said that, in his personal view, the market has rewarded C-level work with A-level money, so that there is no incentive to fix flaws. He laid a portion of attacks on businesses to pranks, often perpetrated by teenagers, involving the misuse of IT tools that have been made available on the Net by their creators. He likened this activity to a practical joke he and his colleagues at a past employer would indulge in: interrupting each other's work by sending a long packet nicknamed the "Ping of Death" that caused a co-worker's screen to come up blue.[30]

The Microsoft OS Source Code Release

The panelists were asked, in view of the release some 10 days before of the Microsoft OS source code, what had been learned about: (a) "security practices in the monoculture"; (b) how this operating system is different from the code of open-source operating systems; and (c) whether Microsoft's product meets the Carnegie Mellon standards on process and metrics.

Saying that he had already seen reports of an exploit based on what had been released, Mr. Walker cautioned that the Microsoft code that had been made public had come through "somebody else." It might not, therefore, have come entirely from Microsoft, and it was not possible to know the levels at which it might have been tampered with. According to some comments, the code is laced with profanity and not necessarily clear; on the other hand, many who might be in a position to "help the greater world by looking at it" were not looking at it for fear of the copyright issues that might arise if they did look at it and ever worked on something related to Windows in the future.[31]

Dr. Lam said that she had heard that some who had seen the code said it was very poorly written, but she added that the Software Engineering Institute processes do not help all that much in establishing the quality of code; there are testing procedures in place, but the problem is very difficult.

Going back to a previous subject, she asserted that there are "very big limitations" as to what can be built using open-source methodologies. A great deal of open-source software now available—including Netscape, Mozilla, and Open Office—was originally built as proprietary software. Open source can be as much an economic as a technical solution, and it is Microsoft's monopoly that has caused sources to be opened.

---

[30]According to some industry experts, most attacks are now criminally motivated, and that the criminal organizations have substantial expertise. They note that the old "curiosity-driven hacker" or "macho hacker" has given way to criminals involved with phishing, bot-nets, data theft, and extortion.

[31]Trade secret laws are an additional concern.

Software Measurement— What Do We Track Today?

INTRODUCTION

Kenneth Flamm
University of Texas at Austin

Dr. Flamm, welcoming the audience back from lunch, said that discussion of software's technical aspects was to give way now to an examination of economic and accounting issues. He introduced Prof. Ernst Berndt of MIT's Sloan School of Management, a pioneer in developing price indexes for high-tech goods, and Dr. Alan White of Boston's Analysis Group, Inc., to jointly report on their analysis of prepackaged software prices.[32]

[32]See Jaison R. Abel, Ernst R. Berndt, and Alan G. White, "Price Indexes for Microsoft's Personal Computer Software Products," NBER Working Paper No. 9966, Cambridge, MA: National Bureau for Economic Research, 2003 and Jaison R. Abel, Ernst R. Berndt, and Cory W. Monroe, "Hedonic Price Indexes for Personal Computer Operating Systems and Productivity Suites," NBER Working Paper No. 10427, Cambridge, MA: National Bureau for Economic Research, 2004.

MEASURING PRICES OF PREPACKAGED SOFTWARE

Alan G. White
Analysis Group, Inc.

Dr. White began by disclosing that Microsoft had funded most of the research on which he and Dr. Berndt had been collaborating, a pair of studies conducted over the previous 4 years. Counsel for Microsoft had retained Dr. Berndt in 2000 to work on price-measurement issues in the software industry, his essential tasks having been to demonstrate how to measure software prices, and to explain how they had been changing over time. Dr. White said that he and Dr. Berndt would not be speaking about the merits or otherwise of Microsoft's actions, but rather would describe their own work in estimating price changes for prepackaged software over time.

Although better estimates of price change existed for prepackaged than for own-account or custom software, Dr. White said, many of those studies were old, dating to the late 1980s or early 1990s. And, in any event, important challenges remained for those constructing measures of price and price change, even when their activity focused on prepackaged software. One such challenge, at the fundamental level, was ascertaining which price to measure, since software products may be sold as full versions or upgrades, as stand-alone applications or suites. Evoking Windows to demonstrate the complexity of this issue, Dr. White ran down a variety of options: buying a full version of Windows 98; upgrading to Windows 98 from Windows 95; or, in the case of a student, buying an academic version of Windows 98. Other product forms existed as well: An enterprise agreement differed somewhat from a standard full version or an upgrade in that it gave the user rights to upgrades over a certain period of time. The investigators had to determine what the unit of output was, how many licenses there were, and which price was actually being measured. Adding to the challenge was the fact that Microsoft sold its products through diverse channels of distribution. It was selling through original equipment manufacturers (OEMs) like Compaq, Dell, and Gateway, which bundled the software with the hardware, but also through distributors like Ingram and Merisel. Prices varied by channel, which also needed to be taken into account. Another issue, to be discussed by Dr. Berndt, was how the quality of software had changed over time and how that should be incorporated into price measures. These issues had to be confronted, because measuring prices matters for producing an accurate measure of inflation, which is used to deflate measures of GDP both at an aggregate level and by sector.

Prices Received by Microsoft Declined Between 1993 and 2001

Dr. White said he would discuss one of two studies he and Dr. Berndt had done, both of which showed that software prices had been declining. The study

Dr. White would summarize used internal Microsoft transaction data and thus was situated "essentially at the first line of distribution," taking into account both primary channels through which Microsoft was selling its products, the OEM channel and the finished-goods or distributor-wholesale channel. The prices he would be referring to would thus be those that Microsoft had received, and whose levels had declined between 1993 and 2001.

In constructing measures of price change, Drs. White and Berndt needed to take into account not only such issues as full versions and upgrades, or academic and non-academic licenses, but also volume license agreements and the shift, which had begun in the 1990s, to selling word processors and spreadsheets as part of a suite rather than as stand-alone applications. In the early 1990s, about 50 percent of word processors were sold as stand-alone components, a percentage that had decreased considerably. Excel and Word were now more commonly sold through the Office suite, with stand-alone sales of the latter dropping to fewer than 10 percent in 2001 from over 50 percent in 1993. Volume licensing sales, representing sales to large organizations in the form of a 500-site license or a 1,000-site license, for example, had grown for Microsoft over time. As to the two channels of distribution through which Microsoft sold, operating systems were sold predominantly through the OEM channel, whereas applications were sold predominantly through distributors.

The study employed matched-model price indexes generally consistent with Bureau of Labor Statistics (BLS) procedures that treated full versions and upgrades as separate products, or separate elementary units, in constructing measures of price change. Dr. White posted a chart demonstrating that price changes varied quite a bit depending on the product, although all Microsoft software product categories posted declines in 1993-2001 for an overall annual average growth rate of minus 4.26 percent during that period (See Figure 10). The rate of decline also varied somewhat within the period studied (See Figure 11). He stressed that the study, based exclusively on prices received by Microsoft, did not necessarily say anything directly about changes in the prices paid by final consumers. In addition, quality change was not explicitly incorporated into its measures of price change, but Dr. Berndt was about to deal with that subject in his talk.

Ernst R. Berndt
MIT Sloan School of Management

Addressing quality change and price measurement in the mail-order channel, Dr. Berndt stated that since the mail-order channel included prices of products that competed with those of Microsoft, a study of it had its advantages over a study limited to the Microsoft transactions data. The disadvantage, however, was that the mail-order channel was becoming increasingly less important, as most current sales were going through the OEM channel and through the resellers or

• Using a matched-model price index to compare 'similar' products over time, prices fallen from 1993 – 2001

Product	Average Annual Change (1993-2001)
Stand-alone Word	-8.34%
Stand-alone Excel	-1.19%
Office	-4.78%
Operating Systems	-0.39%
Allocated Word	-10.64%
Allocated Excel	-8.17%
All Microsoft Products	-4.26%

FIGURE 10 Microsoft's prepackaged software prices have declined at varying rates.

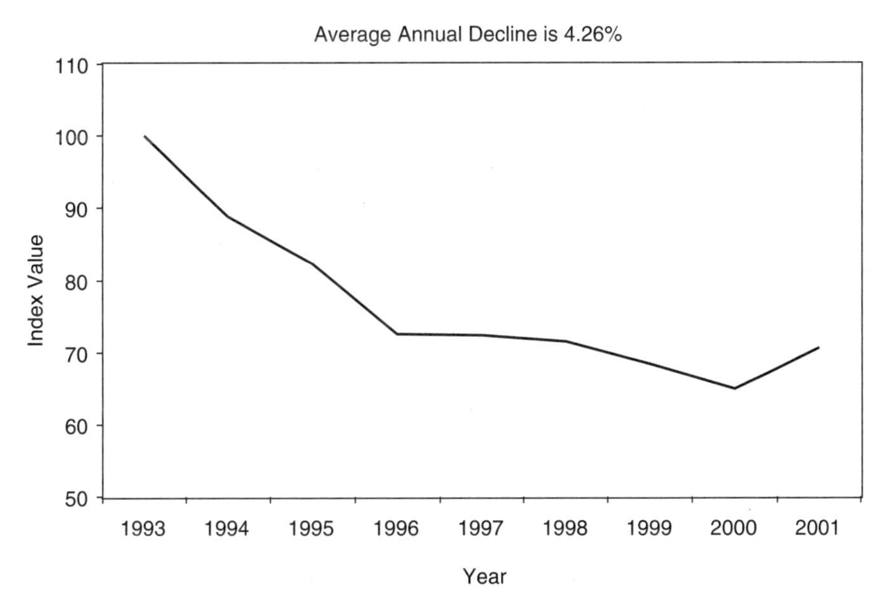

FIGURE 11 Microsoft's prepackaged software prices have declined, 1993-2001.
NOTE: AAGRs are -7.79% (1993-1997), -0.60% (1997-2001), -4.26% (1993-2001).
SOURCE: Jaison R. Abel, Ernst R. Berndt and Alan G. White, "Price Indexes for Microsoft's Personal Computer Software Products," NBER Working Paper No. 9966, Cambridge, MA: National Bureau for Economic Research, 2003.

distributors channel. Drs. Berndt and White had conducted this part of their study for two reasons: (1) because there had been a lot of previous work on the retail channel; and (2) because they had wanted to construct some measures of quality change, or hedonics, for operating systems and productivity suites that, to the best of their knowledge, had not been done before.

Surveying the types of quality changes that might come into consideration, Dr. Berndt pointed to improved graphical user interface and plug-n-play, as well as increased connectivity between, for example, different components of the suite. Greater word length, embedded objects, and other sorts of quality change should be taken into account as well. Hedonic price indexes attempt to adjust for improvements in product quality over time using multivariate regression techniques in which the left-hand variables are prices and the right-hand variables are various measures of quality, and into which time is also incorporated. The product attributes for operating systems had been taken from various documents over the 13-year period between 1987 and 2000; a sample done for productivity suites using prices taken from mail-order ads in the magazine *PC World* covered a longer period, 1984-2000, and also included quality attributes and price measures.

Different Computations, Different Curves

Posting a graph showing the basic results for operating systems, Dr. Berndt explained the three curves plotted on it: "Average Price Level," representing the price per operating system computed as a simple average, which showed an average annual growth rate of roughly 1 percent; "Matched-model," mimicking BLS procedures by using a matched-model price-index methodology, which showed a decline of around 6 percent a year, "a considerably different picture"; and "Hedonic," using a traditional approach of multivariate regressions, which showed a much larger rate of price decline, around 16 percent a year (See Figure 12). Splitting the sample into two periods, 1987-1993 and 1993-2000, highlighted considerable variability in price declines with some more recent acceleration.

For productivity suites, the story was slightly different (See Figure 13). The "Average Price Level" had fallen very sharply in the final few years of the study, in part, because prices for WordPerfect and Lotus suites were slashed beginning around 1997. The "Matched-model" index showed a decline of not quite 15 percent per year with a marked difference between the first and second halves of the sample: zero and minus 27, respectively. "Hedonics" in this case had shown a rate of price decline that was on average a bit larger than that shown by "Matched-model" over the same period.

Recapping the two studies, Dr. Berndt expressed regret at not being able to procure data on the rest of the market, saying that "remains a big hole," but noted that even Microsoft was unable to get data on its competitors' prices. He also pointed to an interesting methodological question arising from the studies: How

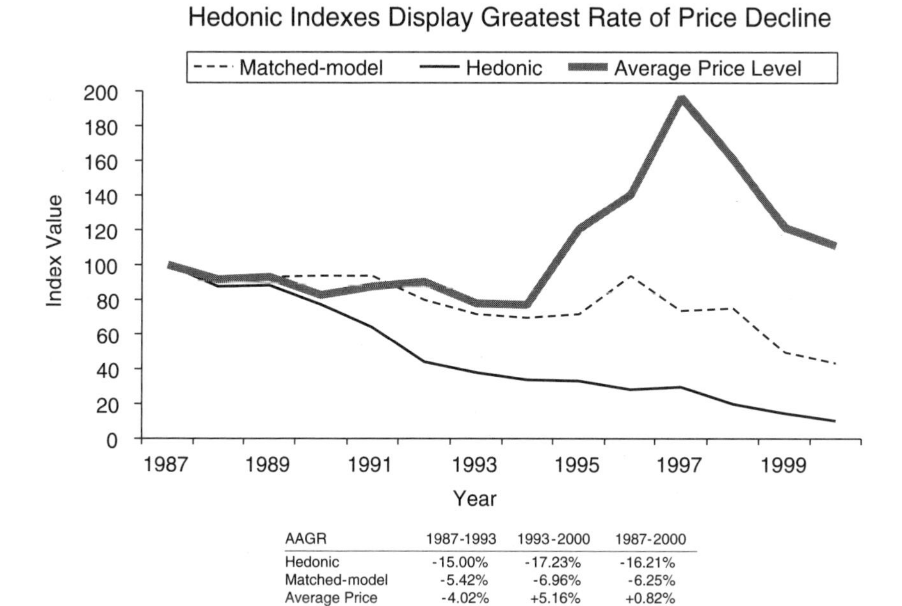

AAGR	1987-1993	1993-2000	1987-2000
Hedonic	-15.00%	-17.23%	-16.21%
Matched-model	-5.42%	-6.96%	-6.25%
Average Price	-4.02%	+5.16%	+0.82%

FIGURE 12 Quality-adjusted prices for operating systems have fallen, 1987-2000.
SOURCE: Alan White, Jaison R. Abel, Ernst R. Berndt, and Cory W. Monroe, "Hedonic Price Indexes for Operating Systems and Productivity Suite PC Software" NBER Working Paper 10427, Cambridge, MA: National Bureau for Economic Research, 2004.

can software price changes be measured and related to consumer-demand theory when software is sold bundled with hardware? The economic theory of bundling was well worked out only for cases in which consumers are very heterogeneous, he stated, adding, "And that's why you bundle." But a price index based on economic theory that is based on heterogeneous consumers raises a number of very difficult measurement issues, as well as theoretical issues.

DISCUSSION

Hugh McElrath of the Office of Naval Intelligence asked Dr. White whether Microsoft had shared its per-unit prices with him or the data had become public in conjunction with a court case.

Dr. White said that he and Dr. Berndt had had access to Microsoft's internal transactions data because it was part of litigation proceedings. He emphasized, however, that their study presented an index based on the per-unit prices they had received but did not disclose actual price levels.

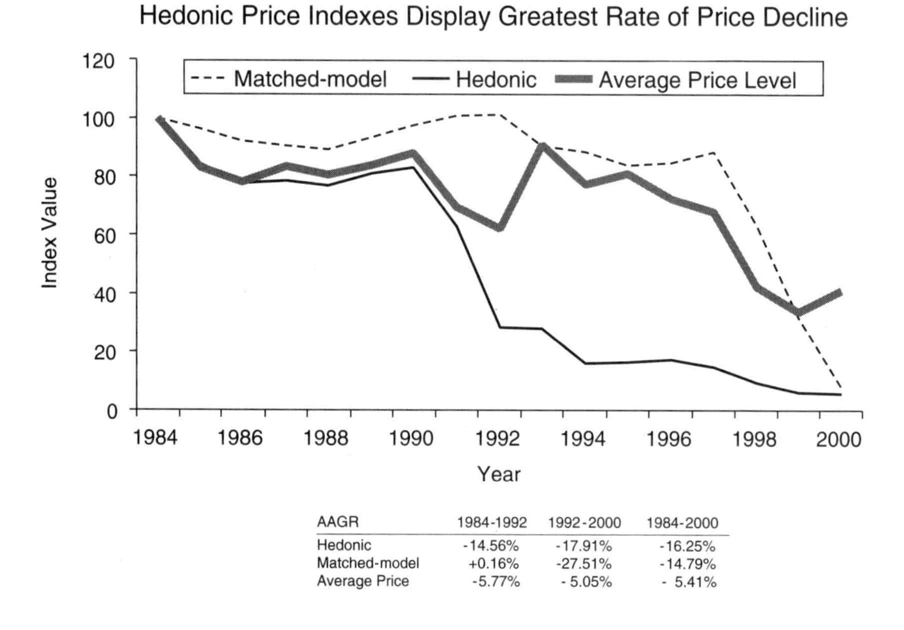

AAGR	1984-1992	1992-2000	1984-2000
Hedonic	-14.56%	-17.91%	-16.25%
Matched-model	+0.16%	-27.51%	-14.79%
Average Price	-5.77%	- 5.05%	- 5.41%

FIGURE 13 Quality-adjusted prices for productivity suites have fallen, 1984-2000.
SOURCE: Alan White, Jaison R. Abel, Ernst R. Berndt, and Cory W. Monroe, "Hedonic Price Indexes for Operating Systems and Productivity Suite PC Software" NBER Working Paper 10427, Cambridge, MA: National Bureau for Economic Research, 2004.

Dr. Flamm pointed to the discrepancies between the MS sales matched-model and hedonic price index results, and the reasons that might be behind them, as an interesting aspect of these two presentations. He asked whether a decline in mail-order margins over time, perhaps with greater competition in the field, could account for them. Second, he wondered whether a matched-model price index could fully capture pricing points between generations of products and speculated that a hedonic index might be able to do so, offering as an example the movement downward of Office-suite prices from one generation to the next. Third, he asked whether it was correct that bundling was mandatory for most U.S. OEMs and, as such, not a decision point, saying he recalled that Microsoft had threatened to sue computer manufacturers if they did not license Windows when they shipped the box.

While Drs. Berndt and White admitted that they could not answer Dr. Flamm's last question with total certainty, Dr. Berndt said that he had been looking at a different question: how to put together a price index that was consistent with consumer-demand theory when bundling is occurring. And he reiterated that the pricing theory on bundling usually put forward was based on heterogeneous consumers.

Dr. Flamm responded that he was only commenting that, in this case, bundling might not have been entirely voluntary on the part of the manufacturers. He then introduced Shelly Luisi, the Senior Associate Chief Accountant in the Office of the Chief Accountant of the U.S. Securities and Exchange Commission (SEC), who was to talk in tandem with Greg Beams of Ernst & Young about accounting rules for software.

ACCOUNTING RULES: WHAT DO THEY CAPTURE AND WHAT ARE THE PROBLEMS?

Shelly C. Luisi
Securities and Exchange Commission

Ms. Luisi said that while she and Mr. Beams would both be addressing the financial reporting that affects software, she would speak more on a conceptual level, while Mr. Beams would speak more about the financial standards specific to software companies and to recognizing software development. Before beginning her talk, she offered the disclaimer that all SEC employees must give when they speak in public, that the views she would express were her own and did not necessarily represent those of the commissioners or other staff at the Commission.

Beginning with a general rundown of the objectives of financial reporting, Ms. Luisi explained that the Financial Accounting Standards Board (FASB) has a set of concept statements that underlie all of its accounting standards and that the Board refers back to these statements and tries to comply with them when promulgating new standards. The Board had determined three objectives for financial reporting:

- furnishing information useful in investment and credit decisions;
- furnishing information useful in assessing cash flow prospects; and
- furnishing information about enterprise resources, claims to those resources, and changes in them.

These objectives stem primarily from the needs of the users of the financial statements, which FASB has defined as investors, whether they are debt investors or equity investors.[33] In light of a general sentiment that financial statements

[33]*Debt investment* is investment in the financing of property or of some endeavor, in which the investor loaning funds does not own the property or endeavor, nor share in its profits. If property is pledged, or mortgaged, as security for the loan, the investor may claim the property to repay the debt if the borrower defaults on payments. *Equity investment* is investment in the ownership of property, in which the investor shares in gains or losses on the property. Definitions of the U.S. Department of Treasury can be accessed at <*http://www.ots.treas.gov/glossary/gloss-d.html*>.

should be all things to all people, it is important to realize when looking at financial statements that the accounting standards used to create them are developed with only one user in mind: the investor. "They are not made for regulators, as much as we might like them to be made for us," Ms. Luisi observed. "They are not made for economists to study. They are not even made for management." It is the goal of providing this one user, the investor, with unbiased, neutral information that shapes accounting standards. The goal is not to influence investors in a given direction or to get public policy implemented in a certain way by making a company appear a certain way; it is purely to present unbiased, neutral information on the basis of which investors can do their own research and determine what decisions they want to make regarding an investment.

Financial statements are part of financial reporting. Disclosures are also part of financial reporting, and they are very important. When promulgating standards, FASB uses disclosures extensively; that a number is not to be found in a financial statement does not mean that the Board has decided it was unimportant. Disclosures are very important from the SEC perspective as well, noted Ms. Luisi, adding, "We obviously have our own requirements in MD&A [Management's Discussion and Analysis] and various other places—in 10-Ks (a type of SEC filing) and registration statements—requiring disclosures that we think are important."

Qualifications for Recognition vs. Disclosure

There are three primary qualifications distinguishing information that must be recognized in a financial statement from information that merely needs to be disclosed. Information that must be recognized:

1. must meet the definition of an element; assets, liabilities, equity, revenue, expenses, gains, and losses are in this category.
2. must trip recognition; an example of an asset that meets the definition of an element but doesn't trip a criterion for recognition is a brand's name. "Surely [Coca-Cola's] brand name is an asset, surely it has probable future economic benefits that they control," acknowledged Ms. Luisi, "but, in our current financial accounting framework, they haven't tripped a recognition criterion that would allow them to recognize that asset on their balance sheet." and
3. must have a relevant attribute that is capable of reasonably reliable measurement or estimate. While historical cost was considered to be such an attribute in the past, the world has been moving more and more toward fair value, defined as "the amount at which an asset (or liability) could be bought (or incurred) or sold (or settled) in a current transaction . . . other than a forced sale or liquidation."

Moving to the terms "asset" and "liability," Ms. Luisi stressed that their definitions and uses in accounting are not the same as in common English or, perhaps, in economics. In its concept statements, the FASB has defined "asset" and "liability" as follows:

- **Asset:** *probable* future economic benefits obtained or *controlled* by a particular entity as a result of *past* transactions or events
- **Liability:** *probable* future sacrifice of economic benefits arising from *present obligations* of a particular entity to transfer assets or provide services to other entities in the future as a result of *past* transactions or events

She stressed that a future economic benefit must be *probable*, it cannot not merely be *expected*, in order to be recorded on a balance sheet as an asset. Additionally, that probable future benefit must be controlled as part of a past transaction; it cannot depend on the action of another party. "You can't say, 'This company has put out a press release and so we know that it is probable that they are going to do something that will result in value to us,' " she explained. "You don't control that benefit—you can't make them follow through."

Tracing how the capitalization (or estimation of value) of software on the balance sheet arrived at its current form, Ms. Luisi recounted that in October 1974 the FASB put out Statement of Financial Accounting Standards No. 2 (FAS 2), *Accounting for Research and Development Costs*. The Board's move to issue this statement the year following its creation indicates that, from the very beginning, it placed a high priority on the matter. This impression is strengthened by the fact that the Board held public hearings in 1973 while deliberating on FAS 2, and the fact that it cited National Science Foundation statistics on R&D in its Basis for Conclusion on the standard. The Board's decision—which predates even its putting in place a definition for an asset—was that R&D was an expense, with the Basis for Conclusion stating that R&D lacks a requisite high degree of certainty about the future benefits to accrue from it.

FASB Rules Software Development to Be R&D

Four months after FAS 2 came out, an interpretation of it, FASB Interpretation No. 6 (FIN 6), was issued. FIN 6, *Applicability of FASB Statement No. 2 to Computer Software*, essentially said that the development of software is R&D also. FIN 6 drew an interesting line between software for sale and software for operations, for which reason different models apply today to (a) software developed to be sold or for use in a process or a product to be sold and (b) software developed for internal use, such as in payroll or administrative systems. Ten years thereafter, in 1985, the Board promulgated FAS 86, *Accounting for the Costs of Computer Software to be Sold, Leased, or Otherwise Marketed*, which Ms. Luisi characterized as "a companion to FAS 2." From FAS 86 came the concept in the accounting literature of "technological feasibility," that point at which a project under development breaks the probability threshold and qualifies as an asset.

FAS 86 thereby gives a little more indication of how to determine when the cost of software development can be capitalized on the balance sheet rather than having to be expensed as R&D.

But 13 more years passed before the promulgation of Statement of Position 98-1 (SOP 98-1), *Accounting for Costs of Computer Software Developed or Obtained for Internal Use*, by a subgroup of the FASB, the Accounting Standards Executive Committee (AcSEC). It was at the recommendation, or request, of the Securities and Exchange Commission's chief accountant in 1994 that SOP 98-1 was added to the AcSEC's agenda and created. During the intervening time, practice had become very diverse. Some companies, analogizing to FAS 86, were reporting their software-design costs as R&D expenses; others, regarding software used internally more as a fixed asset, were capitalizing the costs. SOP 98-1 set a different threshold for capitalization of the cost of software for internal use, one that allows it to begin in the design phase, once the preliminary project stage is completed and a company commits to the project. AcSEC was agreeing, in essence, with companies that thought reaching technological feasibility was not prerequisite to their being in a position to declare the probability that they would realize value from a type of software. It is worth noting that AcSEC's debate on SOP 98-1 extended to the issue of whether software is a tangible or intangible asset. Unable to come to a decision on this point, the committee wrote in its Basis for Conclusion that the question was not important and simply said how to account for it. Ms. Luisi said she believed that, in most financial statements, software is included in property, plant, and equipment rather than in the intangible-assets line and is thus, from an accountant's perspective, a tangible rather than an intangible asset.

Further FASB Projects May Affect Software

At that time, the FASB was working on a number of projects with the potential to affect how software is recognized on the balance sheet:

- **Elements.** With regard to the three qualifications for recognition in financial statements, the Board was going increasingly to an asset/liability model for everything. She noted that "the concepts of an earnings process to recognize revenue are going away," and "the concepts of 'this is a period expense, it needs to be on the income statement' are going away." This represented an important change to the accounting model that most contemporary accountants were taught in school and had been applying, and one that required an adjustment. Internally developed software was recognized on a balance sheet, unlike such intangible assets as a brand name. And, while it was recognized at a historical cost, it had tripped a recognition criterion.

- **Recognition.** With the Internet bubble of the previous decade, when there was a huge gap between market capitalization and equity, the FASB had been under great pressure to explain to investors why there was so much value that was not in the balance sheet. At that time the Board added a project on disclosure of the value of intangible assets, but that had been taken off its agenda about 2 weeks before the present conference; with the pressure apparently off, FASB board members voted 4 to 3 to remove it. According to Ms. Luisi, two of these four believed that investors do not need to know more about intangible assets than was covered by current disclosure, while the other two were at the opposite end of the spectrum: They wanted to go further by instituting a recognition-criteria trigger that would put brand names and other intangible assets on the balance sheet. She called the vote unfortunate and expressed hope that the issue would return to the FASB's agenda in the future.

- **Measurement attribute.** Is historical value or fair value appropriate? In accounting circles, moving to fair value had long been discussed as tantamount to entering "the Promised Land," Ms. Luisi reported, the assumption being that financial statements all would be all fair value one day. But a lively debate had arisen as to whether fair value is truly the most relevant measurement attribute for every single asset and liability on the balance sheet. There were suggestions that the mixed-attribute model, which had always been referred to as a bad thing, might not be so bad—that having some fair-value and some historical-cost elements on the balance sheet might be in the best interest of investors. On the FASB agenda was a project intended to increase discipline regarding how fair value is determined in order to give accountants a little more comfort around that concept. Once accountants become more comfortable, she said, it was possible that the issue of recognizing intangible assets at fair value would be revisited. Such a move was not imminent, however.

At the core of this debate was the trade-off of relevance against reliability. Recalling Dr. Lam's discussion of the trade-off of the completeness of a software product against its features and time to market, Ms. Luisi noted that similar debates regarding trade-offs between relevance and reliability are frequent in accounting. "We can get an extremely relevant set of financial statements with fair values all over them," she said, "but are they reliable? How are the numbers determined? Is some valuation accountant just sitting in a room figuring out what should go on the balance sheet? How is an auditor going to audit it?" Working through these issues is a slow process, but progress was being made.

Before concluding, Ms. Luisi emphasized that the term "fair value" has a meaning specific to accounting that is related to market-participant value and should not be confused with "fair market value" as the IRS or an appraiser might determine it. Even if the accounting model changes so that software is recorded as an asset on a company's balance sheet, it will not carry the value that that company places on it but rather a value that can be substantiated based on the assumptions of many market participants. This means that if the software is so customized that no market participant would purchase it, its balance-sheet value will not be high.

DISCUSSION

Asked by Stephen Merrill of the STEP Board where other countries were on the question of recognizing intangible assets and assigning fair value, Ms. Luisi said that, with regard to fair-value measurements, they were behind the United States, which had more elements at fair value. She noted that a significant debate was in progress, as the International Accounting Standards Board (IASB) was trying to drive European accounting for financial instruments to a fair-value model and the French banks were resisting. She added that international accounting standards were consistent with U.S. accounting standards in not recognizing internally developed intangibles on the balance sheet, which was corroborated by the next speaker, Mr. Greg Beams of Ernst & Young. Most software companies, he added, did not want to have much in the way of capitalized software under their FAS 86-developed products. The window between the moment when such a product hits technological feasibility and the moment when it is out the door and being sold is very narrow, and most of the financial markets are not looking for a big asset that's been created as the product is developed.

Greg Beams
Ernst & Young

To complement Ms. Luisi's overview, Mr. Beams said, he would speak from the perspective of a software company on what triggers the sales numbers as reported in its financial statements and, more importantly, what hurdles it has to overcome in order to be able to recognize revenue. Sales of shrink-wrapped software are generally less complex in terms of when to recognize revenue, whether the software comes bundled with hardware or is purchased separately, and whether the software is ordered through mail order or in a store. In the case of installations of larger enterprise resource planning (ERP) systems, on the other hand, the software often undergoes significant customization, and, depending on the specific facts and circumstances, divergence can often result in how companies report their revenues. But before turning to examples of the latter, Mr. Beams proposed to talk about some of the different models that companies were using to recognize revenue and report it in their financial statements; broader hurdles to revenue recognition as they applied to software companies; and revenue-recognition hurdles that were specific to software companies (See Figure 14).

"From a financial-reporting perspective," he commented, "software-company revenue recognition is one of the more difficult areas to get your arms around as an auditor, and one of the more difficult areas to determine, as a company, exactly

- Overview of Accounting Practices
- General Revenue Recognition Hurdles
  - Evidence of an Arrangement Exists
  - Fees are Collectible
  - Products or Services have been provided
  - Fee is Fixed and Determinable
- Software Revenue Recognition Hurdles
  - Assessing fair value of deliverables
  - Ratable revenue recognition
  - Up front revenue recognition

FIGURE 14 Accounting for software revenue and costs.

how some of the contracts should be recorded." Finally, he would touch on how software users were recording not only their purchases but also internally developed software products, and how these were being reported in the software users' financial statements. He hoped that the discussion would shed some light on the numbers reported by software companies.

Different Models of Recognizing Revenue

Shrink-wrapped software is generally reported at the time of sale and tends to be purchased without additional deliverables such as installation services and maintenance. Most buyers just take the software home and try to install it themselves.

More complex software, in contrast, usually requires some amount of installation, and customers often purchase what is referred to as "maintenance": bug fixes, telephone support, and upgrades. Companies want the software to remain current and so will often pay recurring maintenance fees each year. In moving from a situation where their customers are purchasing shrink-wrapped software to a situation where they are purchasing software with multiple deliverables—maintenance and, perhaps, installation services in addition to the software—software companies come up against accounting rules requiring that they allocate value to each of those deliverables and then recognize revenue in accordance with the requirements for those deliverables.

With software customization, this becomes even more complex: Installing an ERP system, to pick a typical example, can take multiple years, and professional-service fees covering installation and customization generally make up the majority of the overall cost. Testing and customer acceptance may be encompassed in the agreement, and there can be extended payment terms as well.

All of these variables go into the mix, and companies must try to determine how to report revenue for them. Generally speaking, for software installations with significant customization, companies will report revenue on a percentage-of-completion basis under contract accounting, recognizing revenue as the software company performs the customization.

This then results in a wide difference in when revenue is recognized by the software company. As in the examples we just discussed, this varies from a vendor of shrink-wrapped software who recognizes revenue up-front to a significant customization vendor recognizing revenue over time.

All of this revenue is then being reported in the financial statements of the respective software company, and there is additional disclosure surrounding each company's revenue-recognition practices, but at times it can be difficult for the reader of financial statements to understand the revenue numbers that a particular software company is reporting.

Mr. Beams said he would focus next, in discussing both general hurdles and those specific to the software sector, on vendors who were selling software along with implementation and ongoing maintenance but who were involved in customization only in unique circumstances, because it was in that market segment where the most vendors were to be found. The usual aim of such vendors is to develop a product that replicates itself so that they can sell the same product to multiple customers without needing a great deal of horsepower or a lot of "consulting-type folks" providing implementation services, thus generating significant product revenues on more of a fixed cost base.

General Recognition Hurdles as They Apply to Software

Of general market hurdles that apply to this group of software vendors, the first hurdle they must overcome in order to be able to recognize revenues is *securing evidence that an arrangement exists*. Seeking such evidence often means having a contract that was signed before the end of the reporting period. If the contract is not signed before the end of the period, the vendor will delay recognizing revenue until the contract is signed, regardless of (a) whether the customer has paid or (b) whether the customer already had the product and was using it; this is not, however, a problem for most software vendors. The second hurdle involves *whether the products or services have been provided*, which for a software vendor is more typically determined by whether the product has been shipped. As long as the software product has been shipped to the customer before the end of the period, then software vendors have generally been considered to have met this hurdle.

The next two hurdles are a little more difficult to evaluate. Are the fees collectible? Is the customer creditworthy? For a vendor that's selling into a new market or to startup companies, collectibility can become an issue. If the vendor cannot assess *collectibility* at the outset, the company often ends up recognizing revenue on a cash basis—thus, differently than a company selling into more established markets. The last hurdle is even more judgmental: Is the fee fixed and determinable? That is to say, is the software vendor providing extended payment terms or granting *concessions* to its customers? When a software vendor allows a customer to pay through some type of multiple-payment stream, Mr. Beams explained, the customer often comes back toward the end of that payment stream and says: " 'I'm not really getting much in the way of value. The promises that your salesperson made on the front end really aren't materializing. Therefore, I don't think I'm going to make this last payment, and you can take me to court if you want to try to get it.' " The vendor might offer a 20 percent discount in reaction to entice the customer to pay. This is considered a concession, and, if this behavior becomes a pattern, the software vendor can end up with some serious revenue-recognition issues. The software vendor could be obliged to defer all the revenue that it would otherwise recognize at the time the initial sale is made until the contract is completed and until the software vendor has, in essence, proven that no concessions have been made before the revenue can be recognized. Many times, granting such a concession can be a very smart business decision, especially if the sales people in fact did oversell the functionality of the software, but the accounting guidance can become fairly draconian when it comes to the vendor's financial statements and how it then is required to report its revenues.

Hurdles to Revenue Recognition Applying Specifically to Software

Turning to hurdles that apply specifically to software vendors, Mr. Beams said he would next talk exclusively about vendors that provide software licenses and maintenance, and that he would leave aside professional services, the inclusion of which would complicate the revenue-recognition picture considerably.

When more than one deliverable is involved in reporting a software transaction, each of the deliverables must be assigned a value, and revenue must be recognized in association with that separate value. The FASB has defined how fair value is developed for a software vendor in a way that is unique to software accounting; and the Board has indicated that, in order to establish the fair value of a deliverable, the vendor must sell that deliverable on a stand-alone basis. But because these types of software products are most frequently bundled with maintenance, most software vendors in this group have difficulty in ascribing value to individual elements. Also, if the vendor does sell the individual elements separately, it must do so in a consistent range of prices.

In the software industry, however, it is not at all uncommon for vendors to realize 50 percent or more of their sales in the last month of a quarter—in fact, it is not unusual for them to recognize one-third of their sales in the last 2 weeks of a quarter. Customers know that, in order for a software company to record revenue, it must have a signed deal in place before the end of the quarter, so customers will use that timing as leverage to try to strong-arm the software vendor into giving concessions in order to get the best price they can. In these circumstances, developing pricing that is within a consistent range can be more challenging than one might otherwise think.

Warning that he was generalizing, Mr. Beams nonetheless asserted that while "each and every contract that a software vendor would execute has to be looked at individually, generally speaking software vendors in that situation often end up with ratable revenue recognition."

The majority of software companies typically try to get up-front revenue recognition because it gives them a higher growth rate earlier on in their existence, something they believe usually translates into a higher market capitalization. And if a company is considering going public or doing a liquidity transaction, that higher value can translate into more dollars in the company coffers; so it is a lure that, to date, has been difficult for most companies to overcome. While most are chasing up-front revenue recognition, some software companies want ratable revenue recognition and have structured their agreements to get it; moreover, it is not that hard to trip the software accounting rules such that it would be mandatory to record revenue ratably. As a result, of two software companies that are identical in structure and selling very similar products, one may be accounting for its revenue up front, the other ratably. Thus, it is important to understand, in evaluating revenue that is reported in software company financial statements, whether that company is recognizing revenue on an up-front basis or on a ratable basis. And the latter, by so doing, takes away the leverage that its customers might otherwise have at the end of the quarter.

How Purchasers Account for Software Transactions

How software transactions are accounted for by buyers is another source of information on market activity in the industry. Most companies make two types of software purchases, Mr. Beams stated: "They are either buying software and having somebody come in and install it for them, in which case they're cutting checks and recording that as an asset in their balance sheet; or they are developing software internally for internal use and so are capitalizing those costs." He warned that these costs are being put on the balance sheet before going through the profit and loss statements (P&L)—and that they do not go through the P&L until the projects are up and running and actually being utilized, at which point the company would start to record depreciation over some estimated useful life. For this reason, the depreciation associated with the costs of some major projects—projects, he said, that companies were starting to do again, as the economy had become somewhat more robust than it had been for software vendors—probably would not show up for some time. Such projects remained in the capitalization phase as opposed to being in the depreciation stage.

In conclusion, Mr. Beams stated his belief as an auditor that information published in software vendors' financial statements is useful, although mainly to the shareholder. He acknowledged that detail is often lacking, and that distinguishing one software company's reporting from another, and aggregating such information so that it tells a meaningful story, can be extremely challenging.

Dr. Flamm introduced David Wasshausen of the Bureau of Economic Analysis (BEA), who was to speak on how the government makes sense of those aggregated numbers.

A BEA PERSPECTIVE: PRIVATE FIXED SOFTWARE INVESTMENT

David Wasshausen
Bureau of Economic Analysis

Mr. Wasshausen laid out four points he would address in his discussion of how private fixed investment in software is measured in the U.S. National Income and Product Accounts:

1. when software first began to be recognized as capital investment;
2. how estimates of software are measured nominally;
3. how estimates of software prices are measured and what alternatives have been explored recently in an attempt to improve some of the price indexes used; and
4. software trends that have been observed and some of the challenges that those trends present the BEA as it measures private fixed investment in software.

BEA introduced capitalization of software in its 1999 comprehensive revision of the National Income and Product Accounts. Prior to that software had been treated as an intermediate expense, but there are several reasons it should be treated as a capital investment:

- Software exhibits significant growth with key features of investment.
- Software provides a multi-year flow of services.
- Software is depreciated over multiple years.
- BEA was able to eliminate two inconsistencies when it began to recognize software as an investment: (1) before then, only software that was bundled with hardware was being capitalized; and (2) although software has features that

are consistent with other types of investment goods, it was not being treated as such.

Although BEA did not have complete source data to estimate fixed investment in software, something that has not changed in the meantime, it judged that the pros of capitalizing software in the National Income and Product Accounts outweighed the cons, and so it proceeded to do so. It recognized the same three types of software listed by previous speakers: prepackaged, custom, and own-account. Endorsing Dr. Raduchel's earlier statement that software is the single largest asset type in the United States, Mr. Wasshausen placed business' 2003 purchases of capitalized prepackaged software at around $50 billion, those of custom software at almost $60 billion, and those of own-account software at about $75 billion.

This change had less impact on real GDP growth than some had expected, which according to Mr. Wasshausen could probably be attributed to the fact that the price index for software, when it was first introduced as a capitalized good, showed a very gradual increase of about 1 percent. "If our users were expecting it to look more like computer hardware, which has an average annual rate of decline of about 17 percent, then certainly they would be surprised by the lack of impact on real GDP growth rates," he commented. On the other hand, there were those who were somewhat surprised by the sheer magnitude of the numbers; in particular, officials at the Organisation for Economic Co-operation and Development (OECD) were surprised by the large proportion at which BEA was capitalizing and felt that its estimates for intermediate consumption of software were a little low, a subject to which he would return.

Mr. Wasshausen showed a graph juxtaposing the software price indexes of 10 different countries to illustrate that there is no consensus worldwide on how to measure software prices (See Figure 15). He pointed out that software prices in Sweden, Greece, and Finland have been increasing over time, while Australia and Denmark have displayed significant rates of price decline in the same period and the U.S. has been in the middle.

BEA's Methods for Estimating Nominal Fixed Investments in Software

BEA uses a supply-side approach, the "commodity-flow" technique, to measure prepackaged and custom software, starting with total receipts, adding imports, and subtracting exports, which leaves total available domestic supply. From that it subtracts intermediate, household, and government purchases to come up with business investment in software. Demand-based estimates for software available from the U.S. Census Bureau's annual Capital Expenditure Survey for 1998 were quite a bit lower than BEA's estimates; the Census Bureau was working to expand its survey to include own-account software and other information that had not previously been captured. Mr. Wasshausen said it was his understanding that the Census Bureau intended to start publishing this information

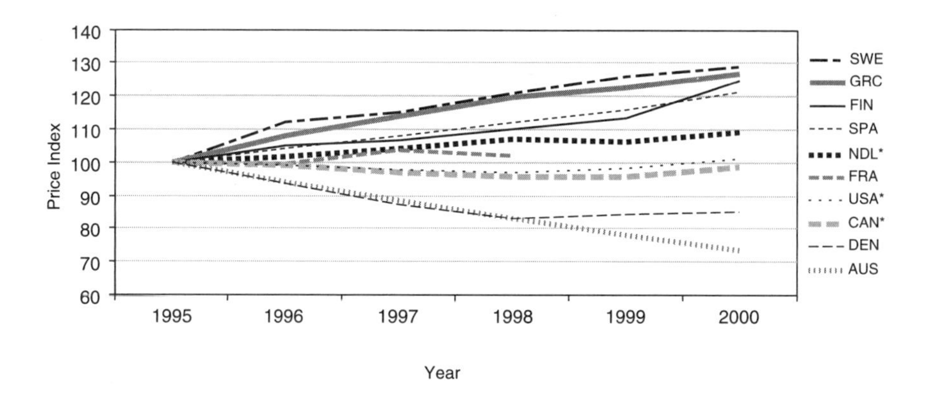

FIGURE 15 Worldwide software prices. Investment in software: price indices from 1995 onwards, 1995 = 100.
SOURCE: Organisation for Economic Co-operation and Development, *Report of the OECD Task Force on Software Measurement in the National Accounts.*

annually as opposed to every five years, the current publication frequency for investment by type. "Having that detailed annual capital expenditure survey data available," he said, "will, if nothing else, provide us with a very valuable check for the supply-side approach."

In contrast to prepackaged and custom software, own-account software is measured as the sum of production costs, including not only compensation for programmers and systems analysts but also such intermediate inputs as overhead, electricity, rent, and office space. BEA's estimates for own-account software are derived primarily from two pieces of source data: (1) employment and mean wages from the Bureau of Labor Statistics' Occupational Employment and Wage Survey, and (2) a ratio of operating expenses to annual payroll from the Census Bureau's Business Expenditures Survey. The computing of private fixed investment in own-account software starts with programmers and systems analysts employment, which is multiplied by a mean wage rate. Several factors are then applied:

- one reduces that number to account for programmers and systems analysts who are producing custom software, in order to avoid double counting;
- another reduction factor accounts for the fact that these programmers and systems analysts do not spend all their work time solely on creating investment software. Much of their time might be spent on things that would not necessarily be characterized as investment, like maintenance and repair;[34] and

[34]For a discussion of the arbitrary assumptions that underlie BEA's estimates of software investment, see Robert P. Parker and Bruce T. Grimm, "Recognition of Business and Government Expenditures

- a ratio that converts from a wage concept to total operating expenses.

If a slide displayed by Dr. Raduchel that put the programmers' cost at only 10 percent or so of the total cost of producing software proved accurate, said Mr. Wasshausen, then BEA was underestimating own-account software.

For the time being, he lamented, BEA had little more to base its quarterly estimates for prepackaged and custom software on than trended earnings data from reports to the SEC by companies that sell such products. "The idea," he explained, "is that as company earnings are increasing, that also means that they must be selling more software." BEA was, however, able to supplement this with data purchased from a local trade source. that tracks monthly retail and corporate sales of prepackaged software. And it expected to have access soon—perhaps from the first quarter of 2004—to better information for the purpose of making the estimates, as the Census Bureau was in the process of conducting a quarterly survey that was to capture receipts for both prepackaged and custom software companies.[35] Meanwhile, BEA's quarterly estimates for own-account software reflected judgmental estimates tied to recent private fixed investment in a variety of areas, including computer hardware, prepackaged and custom software, and some related products.

Recent BEA Software Accounting Improvements

Beginning a rundown of recent BEA improvements, many of them first incorporated in the 1997 input-output accounts that the BEA released in 2003, Mr. Wasshausen pointed to an expansion of the definitions of prepackaged- and custom-software imports and exports used in the aforementioned commodity-flow technique so that the definitions included royalties and license fees, as well as affiliated and unaffiliated software services that are part of BEA's international transactions accounts. Previously, as the BEA was picking up only merchandise in the foreign trade accounts, these were not included. Also improved had been

---

for Software as Investment: Methodology and Quantitative Impacts, 1959-98," paper presented to BEA's Advisory Committee, May 5, 2000, accessed at <*http://www.bea.doc.gov/bea/papers/software.pdf*>. For example, Parker and Grimm note that an important change to the National Income and Products Accounts in the recently released comprehensive benchmark revision is the recognition of business and government expenditures for computer software as investment. Previously, only software embedded in equipment by the producer of that equipment was counted as investment.

[35]Estimates of software based on asking software purchasing companies "how much did you spend?" often yield answers smaller than asking software manufacturers "how much did you sell?" A BEA review of international estimates of software found that countries using the latter question in their approaches typically found much more software investment. BEA also developed alternative purchased software estimates using the Census Bureau's ACES survey of investment, and found that the resulting estimates were about an order of magnitude smaller than the estimates using the Census Bureau's quinquennial economic censuses of software sales, supplemented with its annual surveys.

estimates of how much of the total prepackaged and custom software purchased was for intermediate consumption. This was another area for which the Census Bureau did not collect a great deal of illuminating data, but BEA had augmented the Census Bureau data by looking at companies' reports to the SEC and trying to figure out how much of their revenue came from OEM software sales.

Additionally, BEA was hoping that the Census Bureau would begin, as part of its improved Annual Capital Expenditure Survey coverage, to ask companies how much of their software expenditure was for intermediate and how much was actually a capital expense, and to tabulate the responses. This would allow BEA to harness that information in order to improve its intermediate consumption expenditures for software as well.

In other recent improvements, BEA also had adopted a more direct calculation of total costs for producing own-account software and had replaced median wages with mean wages for its own-account estimates. Finally, it had begun to recognize the production of software originals used for reproduction as capital investment. When a company creates a software original that it intends to reproduce or distribute, it has created an asset. Furthermore, when businesses buy copies of that asset, that is also capital investment. So it is important to capitalize the original production, because what is produced is indeed an asset for the company.

Software Price Indexes

Taking up the issue of price indexes, Mr. Wasshausen noted that the index used for prepackaged software reflected the BLS Producer Price Index and that BEA applied to it a downward bias adjustment of about 3 percent per year. Overall, BEA was "pretty happy" with this index, he said. For own-account software, the Bureau had been using only an input-cost price index that included compensation and intermediate inputs. Because no allowance was made for any changes in productivity, the price index would go up as compensation rates went up for programmers and systems analysts—an obvious weakness in the index. To allow for productivity changes, BEA had begun weighting the input-cost price index, which it assigned a value of 75 percent, together with the prepackaged software price index, which it assigned a value of 25 percent. This had been the same methodology used to compute the custom software price index.

BEA had explored two alternatives for improving its custom and own-account software prices but had yet to implement either. One used "function points," a metric designed to quantify functionality of a given software application to derive a quality-adjusted price index. The idea for the other was to construct a labor-productivity adjustment and to apply that adjustment to the input-cost index.

Function points. At first, BEA had wanted to estimate a price index directly from this function-point data set, whether using hedonics or the average cost per function point. The problem with the function-point data sets that BEA identified was that, while they captured many variables, there was one variable that they did not collect and capture: the true price or the cost of the project. And because it was almost impossible to construct a price index directly from the function-point data set without that information, BEA gave up its efforts to do this. Saying, however, that to his knowledge function points were "the only game in town" when it came to developing a uniform metric for customized software applications, Mr. Wasshausen expressed the wish to be put in touch with someone from the Software Engineering Institute or any other organization with specialized knowledge of software metrics, in the hope he might receive suggestions. "We're certainly open to other areas of improvement," he remarked.

Labor-productivity adjustment. There were two options for creating a labor-productivity adjustment: (1) Using a newly available BLS labor-productivity measure for prepackaged software, and (2) trying to construct a labor-productivity measure directly from the function-point data set. Broadly speaking, BEA would take a relative of the input-cost index and a relative for labor productivity, then dividing the input-cost index relative by the labor-productivity relative to derive a productivity-adjusted input-cost index.

In the case of the first option, the argument was that there was a correlation, or that there were relationships, between own-account software and prepackaged software in terms of productivity changes over time. That both share things like Internet and e-mail, improved hardware, and improved programming tools bolster arguments in favor of using a prepackaged-software labor-productivity adjustment for own-account. But because there are differences as well—for instance, prepackaged software enjoys economies of scale while own-account software does not—BEA agreed to make a layer of downward adjustments to any productivity changes that it got out of prepackaged software, which would create a proxy own-account labor-productivity measure.

In favor of the second option, using function points as the labor-productivity measure, was that these data sets are designed for the task. When companies hire someone to evaluate function points, they are really trying to get a feel for whether a specific software project is on track—or, roughly speaking, how much it should cost. Unfortunately, the results for productivity that BEA obtained using function points showed great volatility compared to the proxy measure constructed using the BLS productivity measure (See Figure 16). The data set it purchased comprised over 2,000 observations spanning 13 or 14 years and captured everything from platform type, the way the function points were actually counted, to development type, whether new development or enhancement was involved. There were many different ways to slice the data, but no matter which way was tried, the results for productivity always came back extremely volatile. Pointing out the sharp drop in productivity in 2001 as measured with the function-point data set, Mr. Wasshausen noted that, from his perspective, the "good news" was that he could argue that the trend was similar between the two sets of data up through 1999 and perhaps 2000.

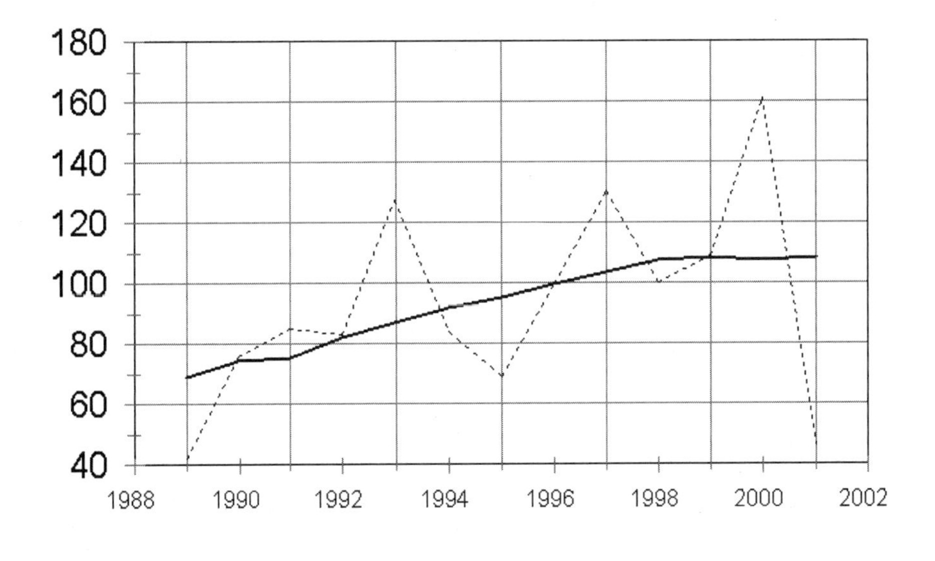

FIGURE 16 Productivity estimates comparison: Own-account SW productivity, 1996 = 100.

Favorable to constructing a labor-productivity adjustment, BEA concluded, were that the issue of including productivity changes needed to be addressed and that the price index that resulted from the adjustment looked more like the price indexes for similar goods. But counterbalancing these "pros" were two powerful "cons": that the adjustments were too arbitrary and that, at the time, making a productivity adjustment was somewhat unorthodox. Believing that the cons simply outweighed the pros, the BEA decided not to move forward.

Software Trends to Present Accounting Challenges

Mr. Wasshausen proceeded to identify and comment upon a number of software trends that BEA had observed:

- **Demand computing.** Suited to businesses with limited software requirements—an example being small companies that use payroll software only once a month—this would allow access on an as-needed basis to an application residing on an external server.
- **Application service providers (ASPs).** Similar to demand computing and seemingly catching on, this practice creates a "very fine line" that poses a chal-

lenging question to BEA: Is software simply being rented, or is an investment involved?

- **Open-source code.** If no dollars are changing hands, it will be very difficult to measure the economic activity that is taking place. While pricing according to the "next-best alternative" may be practicable, Mr. Wasshausen expressed doubt that that would provide any "feel" for overall volume in the economy.
- **Outsourcing overseas.** Looking forward to a session on the subject later in the afternoon, he noted that BEA's International Transactions Accounts measure both affiliated and unaffiliated transactions for such services, which are reflected in its commodity-flow procedure.

In summation, Mr. Wasshausen remarked that, despite its 1999 comprehensive revision, accurate software measurement continued to pose challenges to BEA simply because software is such a rapidly changing field. Characterizing attempts made so far to deal with the issue as "piecemeal"—"we're trying to get the best price index for software, the best price index for hardware, the best price index for LAN equipment or routers, switches, and hubs"—he put forward the notion of a single measure that would capture the hardware, software, and communication equipment making up a system. "If I have a brand new PC with the latest processor and software that works great, but my LAN isn't communicating well with my operating system, I'm going to be processing slow as molasses," he stated. "That's something that, ideally, we'd take into account with an overall type of measure."

Closing, Mr. Wasshausen called the communication taking place at the day's symposium is very important for BEA, as one of its "biggest emphases" is to improve its custom and own-account software prices.

Dr. Flamm then introduced Dirk Pilat, a Senior Economist with responsibilities for work on Productivity, Growth, and Firm-level Analysis at the Directorate for Science, Technology, and Industry of the OECD. Dr. Pilat, said Dr. Flamm, would discuss how the OECD looks at issues like those with which BEA has been grappling in the United States.

WHAT IS IN THE OECD ACCOUNTS AND HOW GOOD IS IT?

Dirk Pilat
Organisation for Economic Co-operation and Development

Dr. Pilat introduced the OECD as an international organization of about 30 member countries whose mission is to promote economic development, in part through improving the measurement of economic activity. He said he would be offering an overview of the official measures of software currently in use for OECD countries, to include measures of software investment; the size of the software industry at a very aggregate level; and software price change. In addition, he would point to various factors that affect the comparability of such measures across countries and would discuss both attempts to improve these measures and the impact improved measures might have on the analysis of economic growth and productivity.

Many of the problems that are inherent in making comparisons across countries had already come up during the symposium:

- **Software is intangible,** and, as such, can be somewhat harder to measure than other products.
- **Markets for software are different from those for other goods,** which means that, particularly as ownership and licensing arrangements are so common, software is a bit more complicated to deal with.
- **Duplication of software is easy and often low cost,** raising the question of whether a copy is an asset. In the current view, the answer was, basically, "yes."
- **The service life of software can be hard to measure,** at least in the way that it is traditionally employed by national accountants.
- **Software's physical characteristics are not always clear.**

These special problems do not invalidate the system of national accounts that is used to measure investment, but figuring out how to apply the system's rules to software does require special effort.

Dr. Pilat displayed a chart based on official OECD data for 18 countries' software investment in 1985, represented by crude estimates; in 1995; and in 2001, the last year for which figures were available (See Figure 17). While observing that, over the period depicted, investment had gone up for all countries, he remarked that there were very large differences among the countries. In Denmark, the United States, and Sweden about 15 percent of total 2001 nonresidential investment had gone to software, but for the UK, a country that might be expected to devote a similar amount of investment to software, official estimates put the total at only about 1.5 percent. "There may be a problem there," he said.

The picture of the computer services industry, which is basically the main producer of software, is somewhat different (See Figure 18). The UK is among the countries with a large industry that produces software services, and Ireland, which was just above the UK at the bottom of the software-investment chart, actually seems to have a large computer services industry. This result again suggests complications that might merit looking at in more detail.

Use of Deflators Also Varies Country to Country

Moving on to the issue of deflators, Dr. Pilat pointed to "very different treatment across countries in how software is looked at," offering as evidence the fact that official statistics for Australia and Denmark showed a very rapid price decline

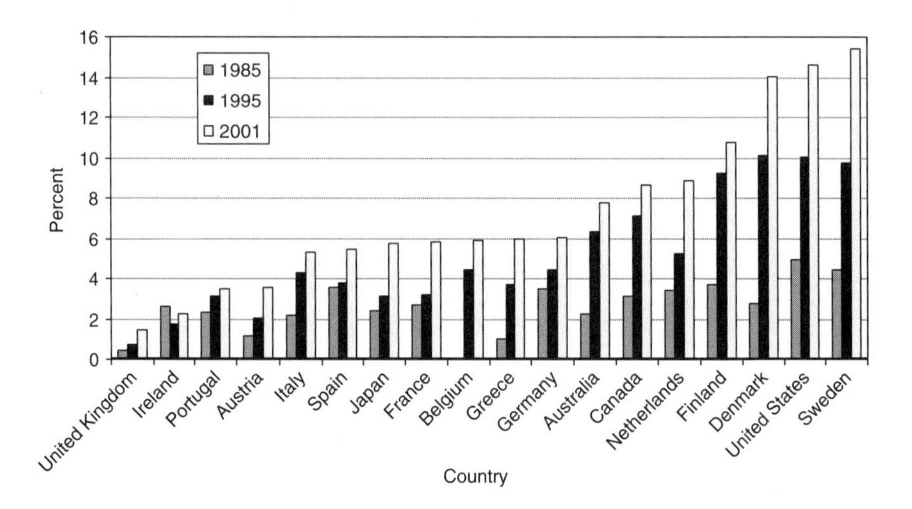

FIGURE 17 The data: software investment, as a percentage of non-residential gross fixed capital formation.
SOURCE: Organisation for Economic Co-operation and Development, Database on Capital Services.

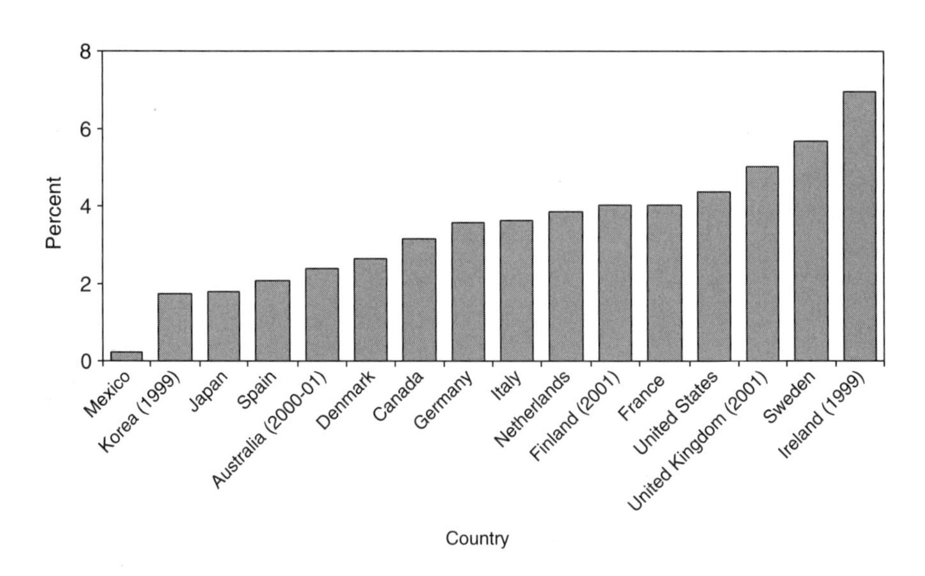

FIGURE 18 Computer service industry as a percentage of total business services value added, 2000.
SOURCE: Organisation for Economic Co-operation and Development, *Technology and Industry Scoreboard*, 2003.

over time, while those for Greece and Sweden showed prices increasing strongly (See Figure 19).

Factors Accounting for the Difference

As one problem contributing to the variation in measures of software investment, Dr. Pilat named that businesses and business surveys—that, he said, generally use "very prudent criteria" when counting software as investment—do not treat software as national accountants might like it to be treated. The consequence is a big difference between business survey data on software investment, which currently exists for only a few countries, and official measures of software investment as they show up in national accounts. Own-account software would not normally be picked up as investment in the business surveys, he remarked.

If business surveys do not reveal much, national accountants must attempt to measure supply using the commodity-flow method described earlier by Mr. Wasshausen. But after ascertaining the total supply of computer services, national accountants make very different decisions on how much of that to treat as investment. Investment ratios therefore differ greatly from country to country, making it quite unlikely that data are comparable For example, about 65 or 70 percent of the total supply of software was being treated as investment by Spain

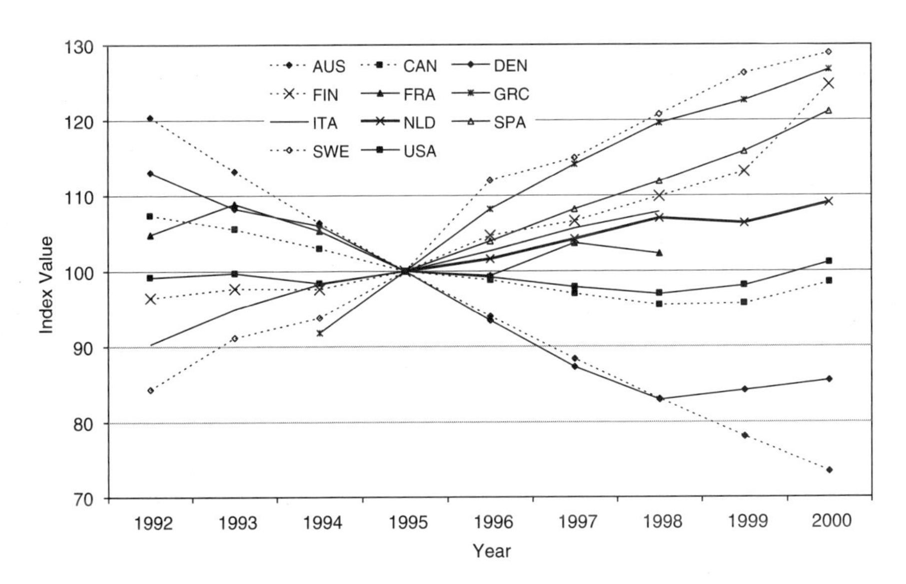

FIGURE 19 Deflators for investment in software, 1995=100.
SOURCE: Ahmad, 2003.

and Greece, whereas the corresponding number for the UK was only about 4 percent (See Figure 20).

What accounts for this difference? It arises in part because the computer services industry represents a fairly heterogeneous range of activities, including not only software production but also such things as consulting services, and national accountants would not want to treat all of them as investment. The main problem is that criteria determining what to capitalize differ across countries. There are also small differences in the definitions of computer services that may exist across countries, although not within the European Union. And there are also problems with accounting for imports, because the trade data don't provide much information on software, as well as with several technical adjustments, which can also differ across countries.

Harmonizing the World Investment Picture Using a Standard Ratio

To some extent, it is possible to tell what would happen if all countries used exactly the same investment ratio. On the basis of an investment ratio for all countries of 0.4 percent—that is, one treating 40 percent of all supply as investment—a very large increase would show up in software investment and GDP levels for the United Kingdom (See Figure 21). Meanwhile, there would be sub-

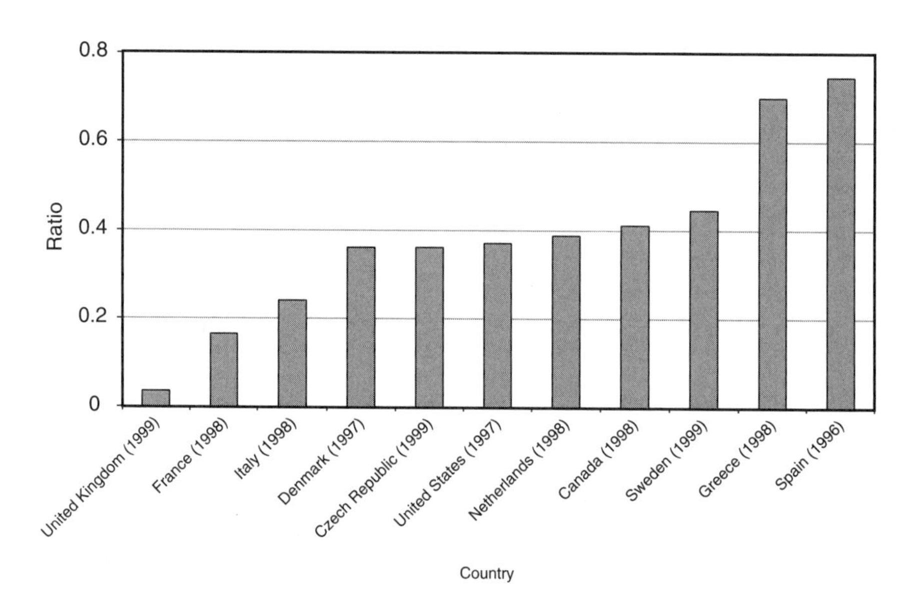

FIGURE 20 Investment ratios for software differ (Share of total supply of computer services capitalized).
SOURCE: Ahmad, 2003.

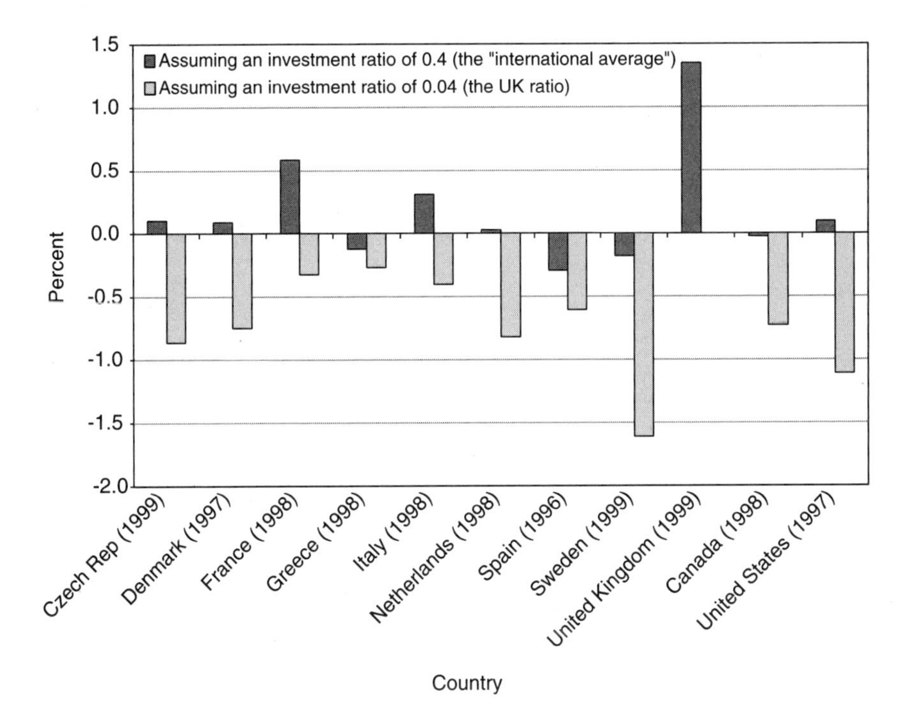

FIGURE 21 Impact on GDP if investment ratios were the same.
SOURCE: Ahmad, 2003.

stantial changes for France and Italy as well, and some decline in a few other countries.

Returning to the problem of own-account software, Dr. Pilat traced the differential treatment it receives across countries:

- Japan excludes own-account from its national accounts altogether.
- Some countries that do include it ignore intermediate costs, looking only at wages and salaries of those who produce it—and then use widely divergent methods of estimating those wages and salaries, especially in regard to the time computer programmers spend on own-account vs. custom production.
- Among countries that take intermediate costs into account, adjustments used for them vary.
- Own-account production of original software designed for reproduction is not capitalized everywhere.

Harmonized estimates reflecting identical treatment of own-account across countries, similar to those for investment ratio already discussed, would show a significant change in levels of software investment (See Figure 22). The portion of Japanese GDP accounted for by own-account software would rise from zero to 0.6 percent, and most other countries would post fairly big increases, with the exception of Denmark, which would register a decrease. Dr. Pilat cautioned that these estimates, which he characterized as "very rough," were not the product of a careful process such as BEA had undertaken for the U.S. economy but had been put together at the OECD solely for the purpose of illustrating what the problems are.

OECD Task Force's Software Accounting Recommendations

In an effort to improve current-price estimates, an OECD-Eurostat Task Force was established in 2001, and it published a report in 2002 that included a range of recommendations on how to use the commodity-flow method, how to use the supply-based method, and how to treat own-account software in different countries. Most of these recommendations had been accepted by OECD countries and were being implemented. Work was also under way to improve business surveys in the hope of deriving more evidence from them over time. If all Task Force recommendations were implemented, Dr. Pilat predicted, the UK would be most significantly affected, but other countries' software-investment data would show rises as well (See Figure 23).

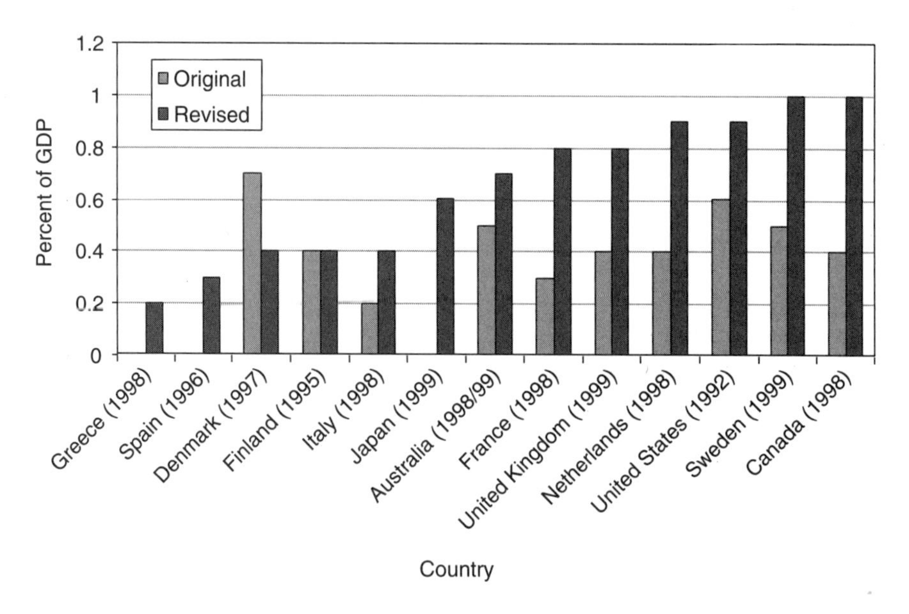

FIGURE 22 Impact of "harmonized" treatment own-account (percent of GDP).
SOURCE: Ahmad, 2003.

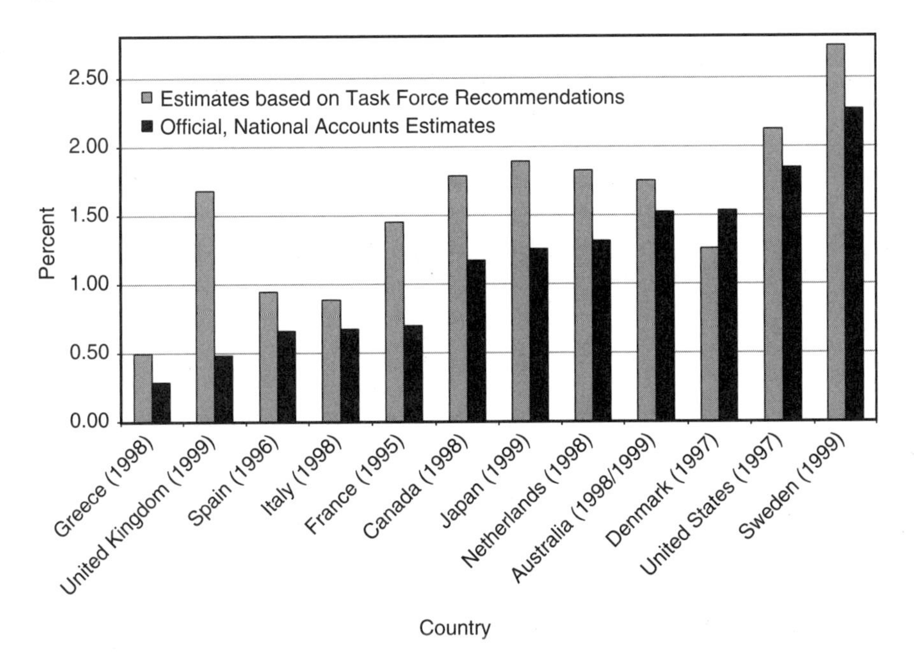

FIGURE 23 Estimated investment in software as a percentage of GDP if Task Force recommendations were implemented.
SOURCE: Ahmad, 2003.

The use of deflators varied widely from country to country as well. While some countries used the U.S. deflator for prepackaged software and adjusted it somewhat, many others used such proxies as a general producer-price index, a price index for office machinery, or input methods. For own-account and customized software, earnings indexes were often being used. The reigning recommendation in these areas, on which the Task Force did not focus to any great extent, was to use the U.S. price index for prepackaged software or to adjust it a little while using earnings indexes for own-account and custom software.

Harmonized Measures' Potential Effect on GDP Levels

As current OECD estimates for information and communication technologies (ICT) investment remained very much based on official measures, adoption of harmonized measures would have the most significant impact on the level of GDP in the UK, France, and Japan (See Figure 24). "While there might be a small change in the growth rate of GDP," Dr. Pilat said, "some [factors] might actually wash out, so it is not entirely clear what that would do to different countries." Software's role in total capital input would definitely increase, which would mean

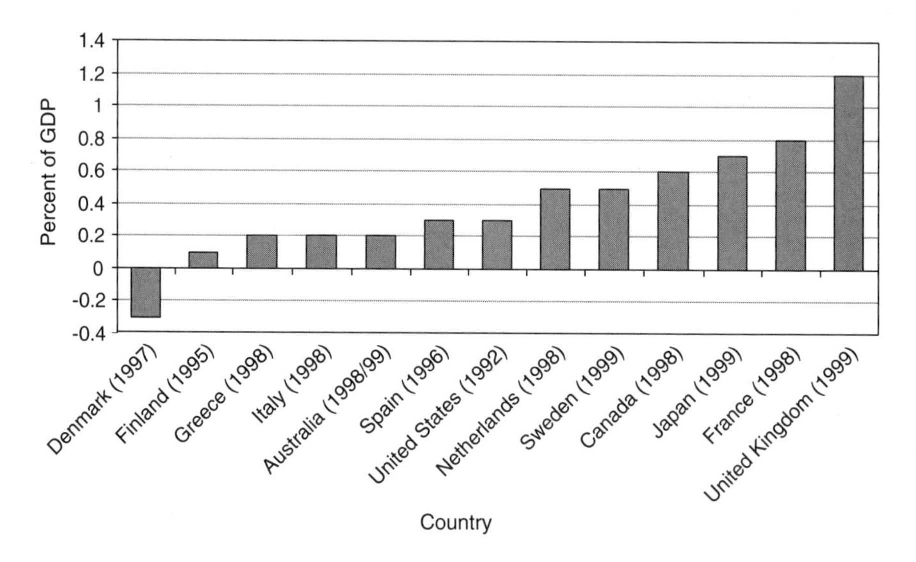

FIGURE 24 Impact of "harmonized" measures on the level of GDP, percentage of GDP.
SOURCE: Ahmad, 2003.

that estimates of multifactor productivity would be changing quite a bit as well. There would probably also be reallocation of other types of capital to software. "Sometimes a country will say, 'we're pretty happy with our total investment expenditure, but we don't quite know where to put it: It may be software, but sometimes we treat it as other types of hardware,'" he explained.

Dr. Pilat then displayed a graph demonstrating that the UK would experience a slight up-tick in cumulative growth rates for the second half of the 1990s if software measurement were changed (See Figure 25). According to another graph, this one showing the contribution of software investment to GDP growth according to growth-accounting analyses for the two halves of the 1990s (See Figure 26), revised estimates would produce a marked increase in that contribution for countries such as Japan and the UK that had a very small contribution to total GDP growth coming from software investment. Contributions of software to total capital in countries like Ireland, the UK, Japan, and Portugal are very small compared to those of other types of IT, which suggests that something isn't entirely right and that a different contribution in the total growth accounting would result from revised estimates (See Figure 27).

Concluding, Dr. Pilat observed that measures of software investment varied quite a lot among countries, but that OECD countries had more or less reached agreement on the treatment of software in their national accounts. Steps were on the way in most countries to move closer to one another in statistical practices,

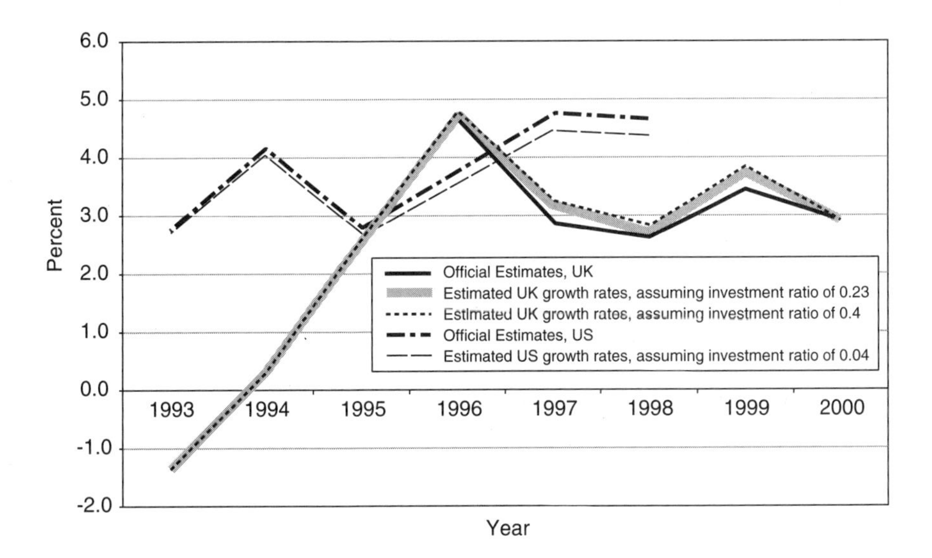

FIGURE 25 Sensitivity of GDP growth to different investment ratios for purchased software.
SOURCE: Ahmad, 2003.

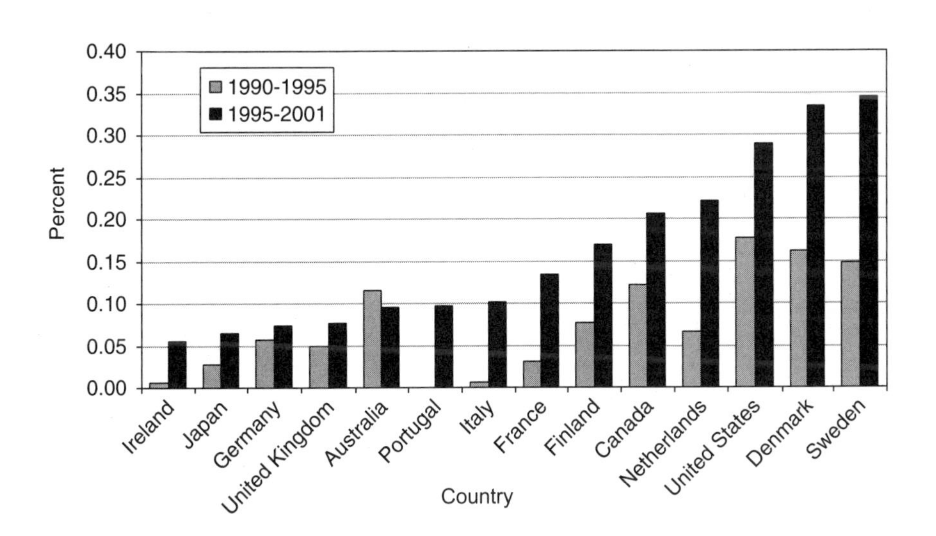

FIGURE 26 Contribution of software investment to GDP growth, 1990-1995 and 1995-2001 (in percentage points).
SOURCE: Organisation for Economic Co-operation and Development, Database on Capital Services, 2004.

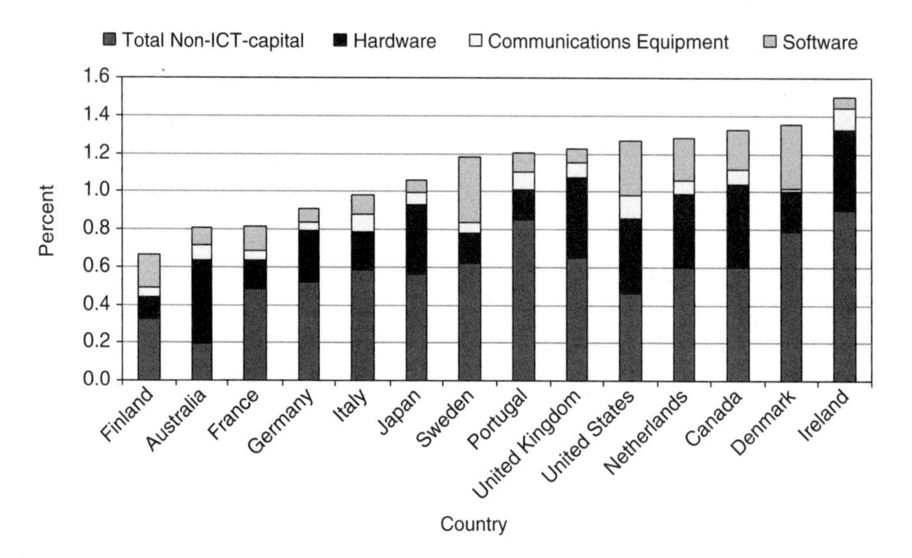

FIGURE 27 Contribution of software investment and other capital to GDP growth, 1995-2001 (in percentage points).
SOURCE: Organisation for Economic Co-operation and Development, Database on Capital Services, 2004.

and there would be some effort by the OECD to monitor the implementation of the Task Force recommendations via a survey due in 2004. But there were some data issues still outstanding, price indexes being one of them. And software trade remained a significant problem, as did detail on software supplies in many important industries, which was not available in many countries. He commended to the audience's attention a working paper entitled "Measuring Investment in Software" by his colleague Nadim Ahmad, which served as the basis for his presentation and was to be found online at *<http://www.oecd.org/sti/working-papers>*.

DISCUSSION

Dr. Jorgenson asked whether it was correct to conclude from Dr. Pilat's presentation that most of the OECD's effort in standardization had been focused on the nominal value of investment, and that the radical differences in prices revealed in one of Dr. Pilat's graphs had yet to be addressed.

Acknowledging that this was indeed correct, Dr. Pilat said that the only place where the OECD had tried to do some harmonization was for the hedonic price indexes for investment in IT hardware. None had yet been done in any software estimates, although he suggested that the OECD could try it to see what it did. He was also considering using both his rough and his revised estimates of software investment to see what effect it might have on the OECD's growth accounting estimates.

Dr. Flamm commented that a lot of interesting work was in progress in all the spheres that had been discussed, and that much work was apparently still to be done at the OECD. He regarded as "good news," however, that improved measures of software prices seemed to be coming down the road, with BEA, university, and other researchers working on the subject.

Moving Offshore: The Software Labor Force and the U.S. Economy

INTRODUCTION

Mark B. Myers
The Wharton School
University of Pennsylvania

Opening the session, Dr. Myers called the attention of the audience to the absence of economists on this panel, which was to address the timely topic of offshore outsourcing. Its speakers, he noted, represented the business and policy communities. He then proceeded to introduce Wayne Rosing of Google, who in his long experience in Silicon Valley had traversed many prominent companies, among them Caere Corporation, Sun Microsystems, and Apple Computer.

HIRING SOFTWARE TALENT

Wayne Rosing
Google

Dr. Rosing took the occasion to present some facts about Google in hopes of dispelling what he alleged to be myths that had grown up around the company during its nearly 6 years of existence. About 40 percent of the company's 1,000-plus employees were software engineers, a figure it wanted to increase to 50 percent; 7 of 11 executive-staff members were engineers or computer scientists, a reflection of Google's identity as an engineering company. Talk of finance and other business topics tended to be crowded out of meetings of the executive staff by its members' focus on "designing things."

Dr. Rosing named four keys to the success that Google had enjoyed to date:

1. **a brilliant idea**—an algorithm called Page Rank that created a better way to organize, rank, and rate sites on the Web—which had been conceived by two Stanford students and out of which had come a better search engine;
2. **a motivating mission statement**, which put forth as the company's aim organizing and presenting all the world's information and trying to make it universally accessible and usable. "That's one of the few corporate mission statements that I believe is actually realizable in, maybe, the next decade," declared Dr. Rosing, who underscored his use of the word "all" in qualifying "the world's information;"
3. **a business model fueling an extraordinary level of reinvestment** in the business, which was very capital intensive, owning large numbers of computers; and
4. **a practice of hiring large numbers of engineers**—"a feedback loop that we keep pushing on."

Dr. Rosing launched his discussion of the panel's topic, which he reformulated as "selective hiring," by stating that Google was "working on some of the hardest problems in computer science" in search of its "Holy Grail[:] . . . that, someday, anyone will be able to ask a question of Google, and we'll give a definitive answer with a great deal of background to back up that answer." This vision, whose fulfillment the company was not predicting for the near future, was carrying it far beyond the technical problem of searching Web pages. In its pursuit, Google had designed and built, and was running, what it believed to be the world's largest distributed computer. At the time of the conference, Google was beginning what was essentially the fourth rewrite of its code, which was constantly being rewritten. The company had two basic hunks of code: (1) Google.com, the familiar search engine, and (2) a set of code for its advertising system and monetization. The company kept the two separate because it had a policy of not selling the ability to get into its index; Google robots had to find a site or information, which was then ranked algorithmically. Employees working on these two code bases were kept apart as well, although they did at times need to interact.

"The Best Minds on the Planet"

The task on which Google was embarked was too difficult to outsource, said Dr. Rosing, adding, "Rather, what we need to do is to pull together all the best minds on the planet and get them working on these problems." Unlike the task itself, attracting the world's best minds to Google was "very simple." In 2003, the company had received a total of 35,000 resumes. As a first step they were culled, mainly by Google engineers but with some recruiters also taking part. From a pool made up of those whose resumes had been selected and prospects who had been referred by employees, the company screened 2,800 applicants. Of those, about 900 were invited to interview, and from that set Google hired around 300, about 35 percent of whom were from the group referred by employees. While the vast majority of Google's employees had graduate degrees from U.S. institutions, they hailed from all over the world, a fact reflected in the makeup of the group of new hires. "So you hire good people," Dr. Rosing observed, and "they know other good people. That's the best source of people."

Of the qualities sought by the company in its hires, Dr. Rosing placed "raw intelligence" first; applicants were likely to be asked their SAT scores, a practice not necessarily common in the corporate world. Second, he named strong computer-science algorithm skills, evidence of which was a degree in computer science from Stanford or the University of California at Berkeley with a superior grade-point average. Google also looked for very well developed engineering skills, as indicated by an applicant having written a program on the order of 10,000 lines in length. "Culture fit" was considered highly important as well, because Google had very little management; 18 months before, Dr. Rosing said, the company had had more than 200 software engineers, and all had worked for him. Really good people, he stated, do not need to be managed, they just need to be given clear goals.

Workers Needed Around the Globe

Google's outstanding problem, Dr. Rosing lamented, was that "there just aren't enough good people." Too few qualified computer science graduates were coming out of schools in North America, including the United States, and in Europe. Short term there was no limit on how many engineers Google was prepared to hire as long as they measured up—and if they could be found. Additionally, pointing out that broadband and other technologies were being deployed faster in many other countries than they were in the United States, he noted that Google's business required having capable employees around the world. Language-specific software skills and, in many cases, a high degree of cultural sensitivity were necessary. The company had begun to hire people abroad, and those people were needed in their own cultures, so it was not practical to bring them to the United States.

Raising a related issue, Dr. Rosing said that Google's in-house immigration specialist had written a memo only days before stating that the H-1B visa quota for the current fiscal year had been filled. Instead of the approximately 225,000 H-1B visas that had been authorized for prior years, the number had been capped at 65,000 in fiscal year 2004—which, he remarked, was an election year. Unable for this reason to hire some people it had in its pipeline to work domestically, Google was opening engineering offices outside the country where foreign employees could be placed after they had been educated, "presumably at taxpayer expense," in the United States. "Now I must ask, why [is this country] doing this?" he said. Google had "smart people, global reach, and scope; and . . . we're intellectually intensive, we're consumed with aggressive reinvention, and we try to automate to the limit to maximize the gross margin per employee." It was, in fact, an example of the type of company that the New Economy was supposed to produce and that would fuel the growth of the nation's economy. Yet the policies in place were limiting the growth of companies such as Google within the nation's borders, something that did not seem to him to make sense.

DISCUSSION

James Socas of the Senate Banking Committee agreed that Google exemplified the company that, kept in the United States, would benefit the country. But he also listed some benefits that the company had derived from the country: employees trained at universities supported by taxpayer dollars; the U.S. regime of property protections; the money of U.S. investors, upon which it would rely when it went public; and the very capitalist system that had given the company life and helped get it through its first 6 years. In light of this, couldn't Google go above and beyond efforts it might otherwise make to hire U.S. engineers who, presumably, would be able to compete with those the company was hiring abroad if given an opportunity on the same footing? Often heard on Capitol Hill, Mr. Socas noted, was: " 'Google and companies like it have benefited so much from this American system, from this American community—is it so much to ask that Google in [its] hiring look first to U.S. workers? What is it that Google owes back to the country?' "

Dr. Rosing answered that, in recognition of the benefit of its having been born in the United States, Google had the responsibility of trying to build a company based in the United States, which he said it was meeting. "We will hire every qualified person we can find in engineering, full stop," he reiterated. "I put no qualification on that. And we can't find enough of them." Saying that the educational system and various factors in society were leading Americans to see other professions as more attractive than science and engineering, he asserted that Google by itself could do little more about this "fundamental problem" than, by going public, create interest and excitement that might inspire others to emulate the company's founders. All Google could do otherwise was to keep hiring and to continue working on the problem of building better search technology.

Asked about attrition at Google, Dr. Rosing placed the rate at well below 1 percent for engineering employees. While speculating that this very low figure might be related to the phase of development in which the company then found itself, he stated that Google's interviewing process was very biased against false positives. "Frankly, we probably miss good people by being so careful," he said, "because it's very difficult to manage people out in the company and, especially when you have very little management, that's a very expensive process." Although the company had added managers, it remained "picky"; avoiding false hires had, in any event, served it well.

Why the Shortage of Qualified U.S. Graduates?

John Sargent of the Department of Commerce noted that the dot-com bubble's collapse had ended the era in which it was difficult to find skilled employees in the information-technology area and, in fact, had left unemployment rates at record highs in almost all IT occupational specialties. If Dr. Rosing was in fact suggesting that the United States was not graduating enough students with skills adequate to meeting Google's needs, would this be because U.S. students were inferior in quality to those trained elsewhere, or because the skills they were being taught in U.S. institutions were incompatible with industry's needs.

According to Dr. Rosing, neither of the two alternatives applied. The United States probably remained one of the world's top areas for computer science education, producing very good graduates. But there weren't enough people going through the system and coming out at the master's and Ph.D. levels to satisfy the needs of the new Information Economy that was the presumed basis for America's competitiveness in the post-industrial world. The fundamental problem—given the excellent quality of the graduates coming out of Stanford, UC-Berkeley, and other U.S. institutions—was that not enough people were going into them. He stressed that he was speaking specifically about computer scientists and not about information technologists in general, a distinction he regarded as subtle but very important.

Dr. Myers then introduced Jack Harding, the Chairman, President, and CEO of eSilicon and a veteran of 20 years of executive management experience in the electronics industry.

CURRENT TRENDS AND IMPLICATIONS: AN INDUSTRY VIEW

Jack Harding
eSilicon Corporation

Mr. Harding introduced his firm, eSilicon, as a three-year-old venture-backed semiconductor company that produced custom chips for its customers in the same way that a general contractor might build a house for an individual. eSilicon implemented its customer's chip specifications by integrating its deep domain expertise with the capabilities of a global, outsourced, tier-one supply chain and delivered a packaged, tested custom chip in volume with the customer's name on it. Among the millions of parts eSilicon had shipped were chips used in the Apple iPod and in a Kodak digital camera. Of the company's 85 employees, about one-third were of Indian descent, one-third of Chinese descent, and one-third of "European or American" descent. As such, he said, "the notion of outsourcing and offshoring is alive and well in our business, and we deal with it in many dimensions every single day."

Mr. Harding said he would frame the problem of outsourcing vs. offshoring from the business person's rather than the economist's point of view, then talk about some of the challenges and trends he and his U.S. competitors faced. He began by citing a recent statement by Autodesk CEO Carol Bartz that he called a "wonderful backdrop" to a discussion of the topic: "When you can get talent at 20 percent of the costs, it isn't about waving the American flag. It's about doing what's right to have a good company." This and a statement by Hewlett-Packard [the then current] CEO Carly Fiorina that "there is no job that is America's God-given right anymore" typified the attitude prevailing in the commercial sector and expressed in various ways by previous speakers: "Hiring here in the United States is important, as is supporting one's nation, but we have businesses to run and that's going to dominate our thinking." This thought process was being "proclaimed throughout Silicon Valley," and echoed, although "more quietly, around the rest of the United States."

Growing Complexity Spurs Outsourcing

The driving force behind outsourcing, as behind other phenomena characterizing the information-technology sector, was complexity. As complexity grows, a firm is forced either to stay ahead of the power curve as long as it can—which means to "sprint like crazy"—or to "step aside and let somebody else do that part of the work." Those areas that are not part of a company's core competency are the first to go. To illustrate, Mr. Harding cited Motorola's decision to spin off its semiconductor business—a business the company had not lost interest in but would have had to recapitalize at a cost of $10 billion in order to stay competitive with outsourcing firms in the industry. "At the point that you're unable to compete by virtue of cost structure or lack of overall efficiency," Mr. Harding stressed, "you are forced to outsource."

Extending this model to the computer industry, he recalled the period some 15 years before when IBM and DEC dominated the PC business as vertically integrated companies. The two rapidly disaggregated across a wide variety of companies to buy memory and processors, to the point that dozens of firms had come to participate in the PC supply chain, whose makeup was changing constantly (See Figure 28). It was as a function of complexity growth that manufacturers needed to find specialists that could fill a particular gap or solve a particular problem. This decision to outsource, he said, was the first step down a "very slippery slope" leading to offshoring. Manufacturers typically began by saying,

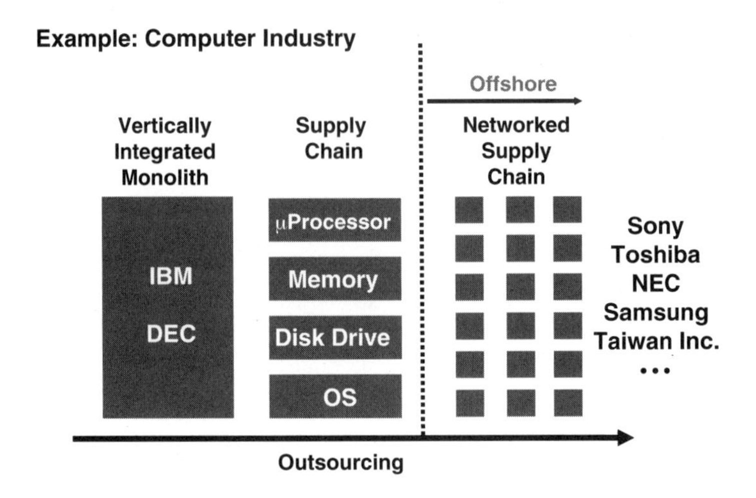

FIGURE 28 Outsourcing leads to offshoring.

"I can't do it myself any longer, and so I'm going to hire somebody nearby so that I can look over them; I really don't want to have them in a different time zone." Upon locating a better supplier "in another county or the next state or the next country," however, they would find themselves embarked on offshoring. Complexity and efficiency working together were, from the business person's perspective, critical to understanding what to outsource and why; when and then where to do so; and whether doing so is worth the challenges that arise with distance.

Displaying an outsourcing/offshoring matrix he said would help illustrate two points central to the ongoing policy debate, Mr. Harding called the audience's attention to the box labeled "captive-offshore" (See Figure 29). That was the locus, he speculated, of "a lot of the political pushback of its being 'un-American' to take jobs to—fill in the blank—Mexico, Taiwan, China." He cautioned those in the policy sector against relying on "a simple formula to understand the 'un-American' aspects of outsourcing or offshoring," emphasizing that specific attributes and market segments merited attention. Activity that could be placed in the "outsource-offshore" box of the matrix, meanwhile, was marked by a trade-off: diminished control against very low variable costs with adequate technical expertise. Recalling the days when "made in Japan" implied questionable quality, he observed that "as you grow up, you realize that also implies things such as IP protection, bootlegged software, [and] cutting the tops off of chips."

	Onshore	Offshore
Outsource	*• Lower Control • High Variable Costs • Peak Load*	*• Lower Control • Low Variable Costs • "Made in China"*
Captive	*• Maximum Control • High Fixed Cost • Traditional Business • Majority Model*	*• Maximum Control • Low Fixed Costs • Large Companies • "Un-American"*

FIGURE 29 *Out/Off* matrix.

Hurdles to Accomplishing Specialized Tasks In-House

Offering an admittedly "simplistic" model that nonetheless might aid understanding of business's outlook on the outsourcing of specialized tasks, Mr. Harding identified general hurdles—such as budget or immigration constraints, or raising the bar too high—that could limit the acquisition of specialized capability in-house (See Figure 30). Specific to companies under $1 billion in size were problems associated with having a specialist on the payroll who could not be kept busy year-round: (a) it was wasteful, and (b) workers of sufficient quality might well reject a position that would not keep them engaged full time. Such factors might incline firms to look outside for expertise needed to get a job done. "There is a fundamental notion," he stated, "that one outsources in order to achieve a specialty skill, regardless of whether it is on- or offshore."

Although the historical reason for going offshore, and the goal implied in the statements of Bartz and Fiorina, was to manage costs, that was not always the motive (See Figure 31). Texas Instruments, for example, had gone to India 20 years before to access a well-educated software pool, which it had managed on a captive basis since. In fact, however, the "vast majority" of the small to medium-sized companies with which Mr. Harding had come into contact were talking about going offshore to save money. Although there might be exceptions to the rule, a software company seeking venture money in Silicon Valley that did not have a plan to base a development team in India would be disqualified as it walked in the door. It would not be seen as competitive if its intention was to hire

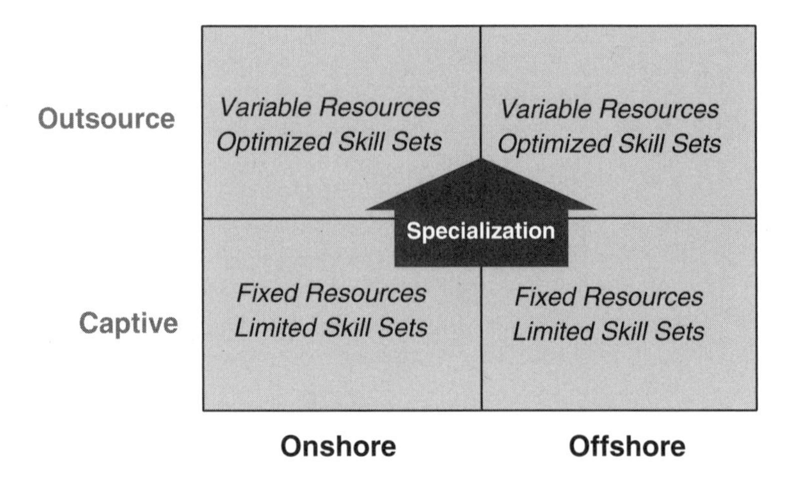

FIGURE 30 Business drivers: outsourcing.

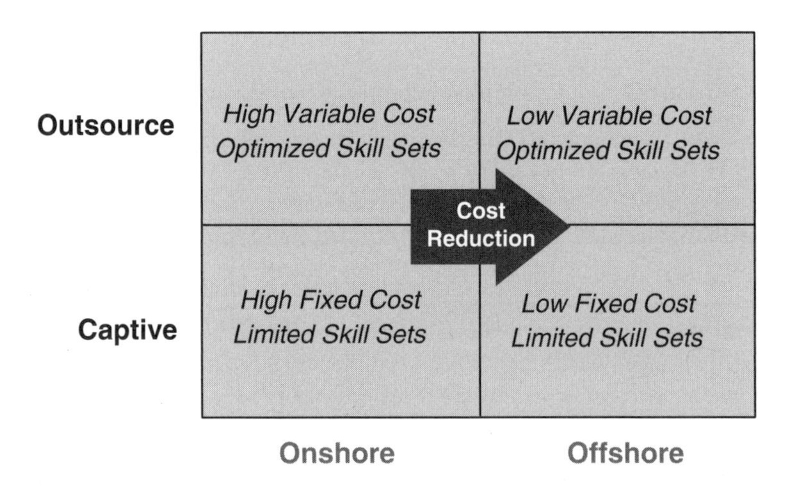

FIGURE 31 Business drivers: offshoring.

workers at $125,000 a year in Silicon Valley when comparable workers were available at $25,000 a year in Bangalore.

Morgan Stanley had predicted that the outsourcing market overall would grow to approximately one-third of the $360 billion semiconductor industry by the end of the decade, a big jump from around one-seventh of a $140 billion market in 2002 (See Figure 32). To suggest the role of offshoring in this change,

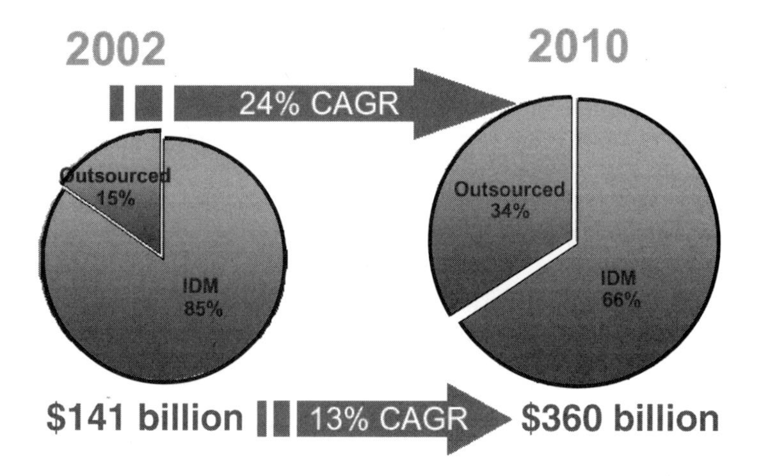

FIGURE 32 Outsourcing trend: semiconductor revenue by source.
SOURCE: Morgan Stanley Research.

Mr. Harding showed a graph tracing the growth of India's software industry from almost nothing to $10 billion in around the same number of years and noted that it had shown no signs of slowing down. The strength of the offshoring trend, he said, made imperative that the United States find a way to become comfortable with it, both from a business and from a public-policy perspective.

Moving Software Development Offshore

Mr. Harding then showed a matrix whose purpose was to organize discussion of the specific topic of moving software development offshore (See Figure 33). Starting with the quadrant defined as "offshore-captive," he pointed out that placing only large enterprises there most likely represented an inaccuracy. While historically it had been companies like TI, Motorola, and Microsoft that hired software developers in India or elsewhere abroad, having a foreign presence had become incumbent even on startups seeking venture capital, as mentioned above. He said that every company he knew of, without exception, was in the process of moving software development to some degree to the Indian marketplace, and that it was inconceivable that any firm would rule such a move out. Although eSilicon itself, a hardware developer that provided its customers with hardware design from time to time, did not yet have an offshore facility up and running; it had made the decision only 2 weeks before to start one. "I consider myself a laggard in that respect," he said. "I'm looking around and saying, 'I'm the last guy on the block to do this.' "

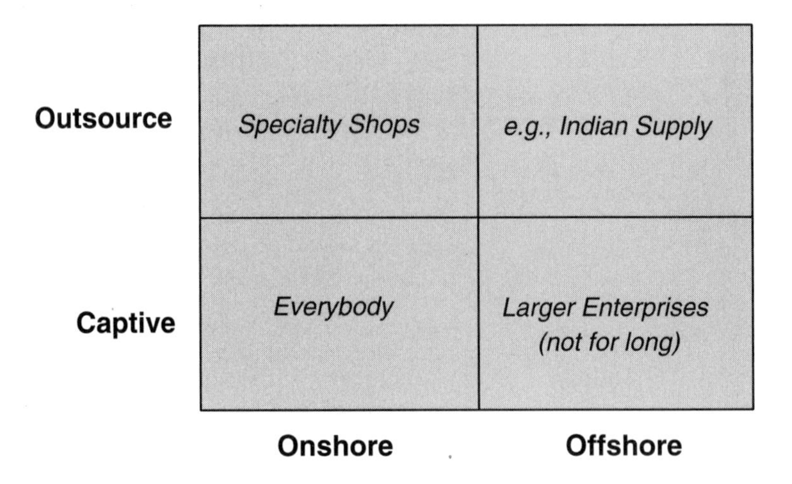

FIGURE 33 Offshoring software.

Offshore-outsourced software development worked first and foremost because of low costs: While costs were increasing compared to previous years, it was still cheaper per person to build software abroad, even taking productivity issues into account. There was a large supply of graduates outside the United States, and the number of graduates at home had diminished to the point that it was difficult to hire good people. Mr. Harding agreed with Dr. Rosing that neither U.S. education nor the students it produced were at fault, but that with the number of graduates down employers were forced by competition for their services into paying them too much. "You're paying master's-level grads out of a fine school around $100,000 to come in and be an apprentice, essentially, for the first 3 years," he commented. "That's not a sustainable model."

The other key factor in favor of offshore-outsourced software development was that the tools involved were, effectively, a commodity. They were robust, had a broad user base and well-developed support infrastructure, and benefited from strong standards. This was significant because, with each advancement of a software product, 18 percent of functionality was lost owing to inability to manage the bugs associated with it. From his days as head of a large software company, Mr. Harding recalled that every time the firm released a new product, it enabled new bugs even as it fixed bugs that were on its customers' top-ten lists. "It's a never-ending process," he lamented. Hence the advantage of having the tools that, while not perfect, was ubiquitous and could be operated effectively by millions of people.

Mr. Harding then paused to offer a comparison between offshore outsourcing of software and offshore outsourcing of hardware that left the former looking in

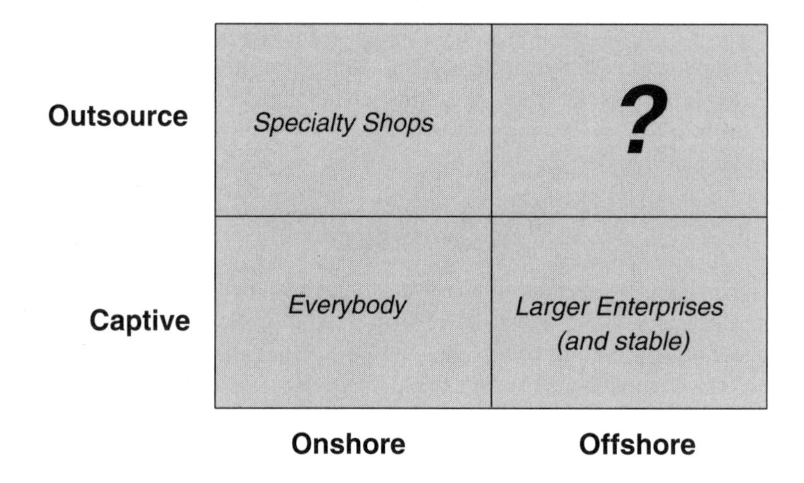

FIGURE 34 Offshoring hardware.

better shape. Drawing a contrast to both the matrix quadrant for software "offshore-outsource" and that for hardware "offshore-captive"—the latter housing any of "25 first-rate electronics firms that do this very successfully"—he argued that no market of any major account yet existed for offshore-outsourced chip hardware (See Figure 34). The reason was that, unlike enterprise software tools, those for chip hardware were not ubiquitous and needed more technical support to make them work. They had a limited user base of tens of thousands of people maximum against millions for enterprise software. In addition, the farther production strayed from Silicon Valley, the hub of design-automation software, the more difficult it was to get the support that was needed to repair a bug, get a patch, or solve whatever other class of problem arose.

Policy Issues: Job Redistribution, Education

Moving to policy issues, Mr. Harding first mentioned loss or redistribution of jobs, then education: Would people come to the United States, get an education, and leave? Or would they not come at all, because the market had shifted to other zones? A third issue was U.S. access to foreign markets—and not only to customers' dollars that might be earned, but also to the R&D and intelligence residing there. All three issues involved risk from a policy perspective, he warned.

In conclusion, Mr. Harding called the outsourcing and offshoring trends "irreversible" and said they needed to be dealt with. "We're going to have to be articulate and thoughtful about all of these points," he stated, "and also respectful of both sides of the argument." He boiled down the debate over offshore outsourcing to the single question that, he said, was likely to be the most relevant to those in attendance—and also most likely, somewhat unfortunately, to occupy politicians' field of vision: Are we trading jobs for earnings per share? It was the task of the business community, he said, to teach its constituencies that contrasting models can coexist.

DISCUSSION

Asked when he expected Indian firms to begin taking his own market share, Mr. Harding replied that his particular firm was unlikely to be at risk because it aggregated many elements of a complex supply chain, such as wafer production from Taiwan and packaging from Korea. As a very small percentage of its business was actual design—of either software or hardware—it accessed the market directly. "There is no one company in India that would have the advantage over us," he said. "They still have to buy from the same suppliers from whom we buy."

The IT Sector's Judgment Questioned

David Longstreet of Software Metrics said that what he had observed through working with companies to put measurements in place prior to outsourcing agreements ran contrary to what Mr. Harding had been suggesting. "When we have measurements in place before and after," he claimed, "what we see is lower quality and lower productivity like for like." Recalling the IT sector's recent missteps—"we were wrong about Y2K and we all rushed to that, we were wrong about dot-coms and we all rushed to that"—he recommended that the industry "learn to tread lightly" and get better measurements in place before rushing to India to outsource software. Warning that the industry was again "rushing to something that is actually drawing a lot of companies to lower earnings," he asked: "Why should an investor believe IT now, when we've been wrong the last two major times?"

When he had been in the software sector, Mr. Harding countered, he had had 300 or 400 employees in India producing software of the same quality as that coming out of Silicon Valley. Hardware was a different story, however, prominent among the reasons being "the availability and application of complex design tools that break easily."

Kenneth Walker of SonicWALL stated his disagreement with Mr. Harding's point on hardware outsourcing, which, he said, was mainly based on the latter's limiting the definition of hardware to chips. Design and production of such hardware systems as TiVo devices, TV sets, and monitors were being outsourced to Taiwan and other countries in large quantity, he noted.

Acknowledging Mr. Walker's comment as fair, Mr. Harding nonetheless argued that the assembly of the items Mr. Walker had mentioned differed in respect to the percentage of the cost of goods sold from the development of the semiconductor components enabling those particular items. System-final assembly certainly was dominant in Asia, which, he added, was the right place to do it.

Dr. Myers then introduced Ronil Hira, a member of the Public Policy faculty at Rochester Institute of Technology (RIT) who specialized in workforce issues and innovation policy.

IMPLICATIONS OF OFFSHORING AND NATIONAL POLICY

Ronil Hira
Rochester Institute of Technology

Dr. Hira specified that he was addressing the conference in two capacities, one as a representative of RIT, the other as chair of the Career and Workforce Policy Committee of the IEEE-USA. The committee, which represents the 235,000 U.S. members of the IEEE, itself a transnational organization, was started in 1972 in response to poor labor markets for engineers early in that decade.

Dr. Hira began by offering a set of definitions:

- **Outsourcing:** a classic make-or-buy decision between producing in-house or purchasing from a supplier. *Example*: Procter & Gamble contracting with Hewlett-Packard for information-technology services.
- **Offshore outsourcing:** using a supplier operating abroad. *Example*: HP's recently reported attempts to "move as much [work for clients] offshore as they can."
- **Offshore sourcing**, or **offshoring:** a multinational corporation that operates offshore. *Example*: Daimler Chrysler's R&D center in Bangalore.
- **Onsite offshore outsourcing:** Bringing foreign workers into the U.S. to work on a project onsite, often on an H-1B or L-1 visa. *Example*: Tata Consultancy Services, Infosys, Cognizant, Accenture, Wipro, or Satyam providing foreign nationals to work in U.S. domestic operations.

Prominent on the list of factors encouraging U.S. companies' use of overseas technology workers were lower labor cost; access to exceptional talent; access to local markets; tax holidays and other industrial-policy measures by countries that are targeting these jobs and industries; and the option to operate around the clock. According to Dr. Hira, however, the strongest motivation had been that companies had become aware of this possibility and believed that taking advantage of it would help their performance. "They're acting rationally," he said, noting that firms follow whatever course permitted by existing rules "they think is in the best interest of the shareholders and management." For this reason, it was his practice to refrain during policy discussions from making appeals to companies' patriotism or moral character; doing so, he said, "makes no sense."

Do Foreign Engineers Truly Work for Less?

Taking up the first entry on the list, labor cost, Dr. Hira displayed a chart comparing salaries for engineers in the United States and several other countries (See Figure 35). Using data expressed in purchasing power parity strongly suggested that, given the variation in the cost of living among nations, it was a misconception that workers outside the United States were willing to work for less. In fact, these workers could afford to be paid significantly less than workers in the United States because, even at lower wages, they would be just as well off. He nonetheless cautioned against the capacity of the purchasing power parity measure to reflect economic reality, noting that in India, he was able to buy a tomato for the equivalent of two U.S. cents, but housing is relatively more expensive in India. He disagreed with those who expected prices to rise very rapidly in India, positing that "there's kind of a governor on that effect," as well as with those who saw wages going up quickly as demand for labor increased. "There's a lot of labor [available] there," he observed.

That no one could say how much software work actually had moved offshore to that point was called a "major problem" by Dr. Hira. There was no one in the federal government collecting data, although a pilot study of the question had begun in the Commerce Department. That effort, budgeted at only $300,000, would of necessity be a modest one and perhaps of most interest to those firms already outsourcing offshore; he expressed the opinion that, if it was to be addressed at all, the issue should be addressed more seriously. But getting an accurate picture was difficult, as two concerns had made companies quite reluctant to reveal their plans in this regard: (1) that they would take a public-relations

Country	Purchasing Power Parity (PPP)	Salary
U.S.	1.0 * $70,000	$70,000
Hungary	0.367 * $70,000	$25,690
China	0.216 * $70,000	$15,120
Russia	0.206 * $70,000	$14,420
India	0.194 * $70,000	$13,580

FIGURE 35 Overseas software engineers *can afford to be paid less.*

hit, and (2) that they would face workforce backlash as their employees found themselves effectively training their own replacements as part of what was euphemistically known as "knowledge acquisition" or "knowledge transfer." The easiest way to find out what was happening was to read in online editions of Indian newspapers the numerous announcements of major U.S. corporations expanding their activities there and the financial statements of the major IT offshore outsourcers like Cognizant, Infosys, and Wipro.

Higher-Level Jobs Have Joined the Exodus

A common misconception, although one not shared by the day's previous speakers, was that only low-level jobs were moving offshore. Evidence to the contrary had been provided by the *Detroit Free Press*, which had reported not long before that General Motors was moving some of its R&D abroad. DaimlerChrysler had R&D operations in Bangalore, and Texas Instruments had been there for some time. But the evidence suggesting that higher level jobs were involved was largely anecdotal, we don't know the true scale and scope, so honest discussion of the situation would have to await more scientific data. Meanwhile, estimates of the number of jobs expected to go offshore had come from firms like Forrester Research that were self-interested in that, as suppliers of advice and other services to firms considering the offshoring option, they stood to benefit from promoting the idea. Furthermore, in Forrester's case the report had been released over a year before, but its numbers were still frequently cited despite the fact that business strategies had changed significantly in the interim.

Dr. Hira displayed a breakdown of employment totals and jobless rates in the U.S. domestic IT labor market, which he said was experiencing "record unemployment" (See Figure 36). He cited a frequent reaction to this table: "That's not a big deal, it's pretty close to the general unemployment rate." But his next graphic, depicting unemployment rates in the technology sector over the previous two decades, provided a perspective (See Figure 37). While the general civilian unemployment rate had trended downward between 1983 and 2003, the rate for computer scientists, after staying in the 1-3 percent range from 1983 to 2000, had climbed to over 5 percent by 2003. The rate for electrical engineers had actually fared even worse over the previous year, surpassing the general unemployment rate in 2003 for the first time in the several decades during which such data had been collected. Nevertheless, over the preceding year the SOX index, which measures the performance of semiconductor-industry stocks, had doubled. "So maybe there's some hope," he speculated. "We'll see when that hiring starts to pick up."

Offshoring's Contrasting Long- and Short-Run Impacts

A change in the mix of U.S. domestic occupations, with attendant job dislocation, was one consequence of the departure of technology jobs overseas that

Occupation	Employed (Thousands)	2003 Unemployment (Percent)
All Managers	14,468	2.9
Computer & Information Systems Mgrs	347	5.0
Engineering Managers	77	3.6
Computer Scientists & Sys Analysts	722	5.2
Computer Software Engineers	758	5.2
Computer Programmers	563	6.4
Computer Support Specialists	330	5.4
Computer Hardware Engineers	99	7.0
Electrical & Electronics Engineers	363	6.2

FIGURE 36 Domestic IT labor market: *record unemployment.*
SOURCE: IEEE-USA

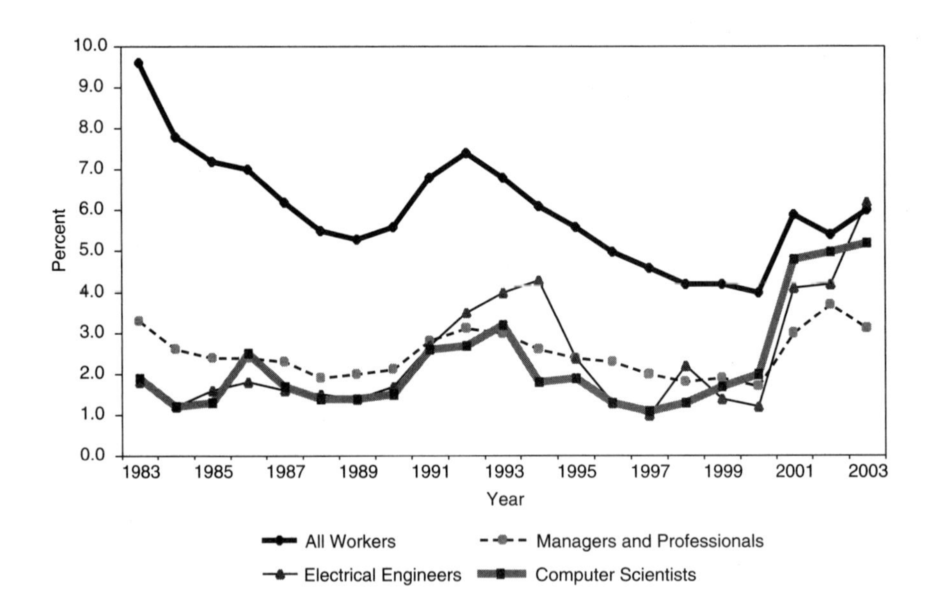

FIGURE 37 1983-2003 tech unemployment rates.

most observers were predicting. While some hoped offshoring would make the U.S. IT job market stronger in the long run, the opposite impact was expected for the short run. Just that week, in fact, Siemens had announced the relocation of almost all the software developers it had been employing in the United States and Western Europe—a total of 15,000—to Eastern Europe, India, and China. "At least they've announced it publicly," Dr. Hira declared. "A lot of people are probably planning it and not saying it." Such moves were apt to prolong unemployment duration in an IT labor market that was weak already, as well as exercise downward pressure on wages. *BusinessWeek* had recently reported that an employer located in Boston placed an ad to hire a senior software engineer at $40,000 just as a lark and received about 90 responses from highly qualified senior software engineers, who were normally paid twice that. Such downward wage pressure had been characterized as the silver lining of the movement offshore by an IT industry representative, Harris Miller, who argued that forcing down wages at home would mean that the United States could keep more jobs. A business magazine columnist, perhaps similarly, had voiced surprise that anger over offshoring was real. The reason for the anger, Dr. Hira suggested, was that "unlike what some are saying—'your job is not moving to Bangalore'—for a lot of software people in fact it is, at least right now."

As to the decade ahead, he displayed a chart showing that the Bureau of Labor Statistics (BLS) had just revised downward by nearly 1 million jobs its forecast for overall job creation in computer-related occupations (See Figure 38).

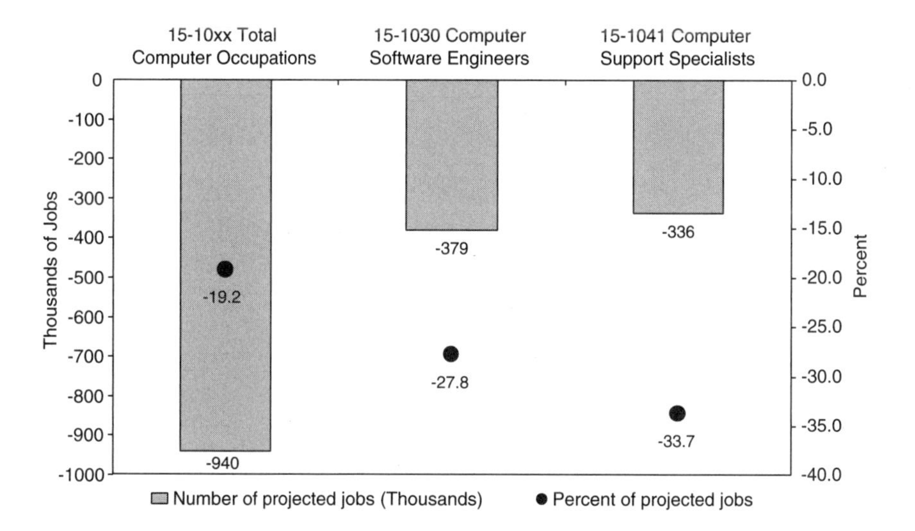

FIGURE 38 Occupation mix—revised BLS job projections: Decrease in number of jobs forecasted, 2002–2012 projections vs. 2000-2010.
SOURCE: Bureau of Labor Statistics.

In updating its projections to run through 2012, the BLS had predicted that, instead of growing from 3 million to 5 million, this job category would grow to only 4 million. While expressing skepticism as to how accurately the occupation mix could be predicted 10 years out, Dr. Hira explained that the chart was a means of countering assertions that the nation would be creating large numbers of "great jobs" in IT and so there was nothing for the sector's employees to worry about. "I hope that the people who are using that as an argument start to update at least some of that data," he remarked. A second chart, this one tracking job creation from 1984 to 2003, showed that the United States' job-creation engine had stalled beginning around 2000 (See Figure 39); even if the general trend resumed in the long run, in the short run employees were being displaced. "They're not going to the IT occupations, where jobs are not going to be created," Dr. Hira stated. "Are they going to go somewhere else? To what professions?"

He raised a number of questions regarding the longer-term impacts off-shoring might have on U.S. innovation:

- Was it important to maintain a strong software and engineering workforce in the United States?

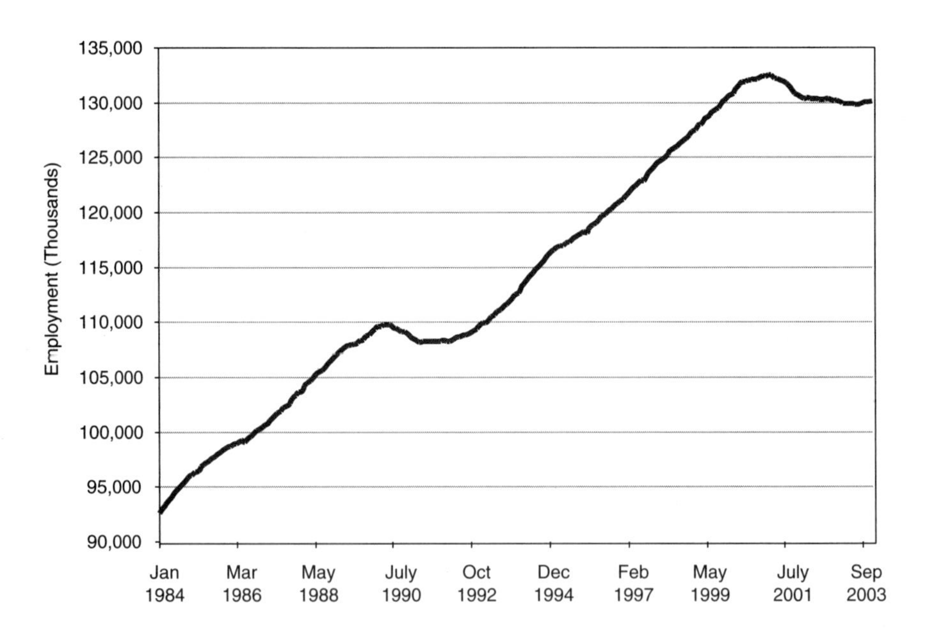

FIGURE 39 Job dislocation during low job creation: Total nonfarm payroll employment (seasonally adjusted).
SOURCE: Bureau of Labor Statistics.

- What would the country's new occupation mix be if software were lost?
- Would losing software have a chilling effect on people's pursuing IT occupations and disciplines?
- Where would the future's technology leaders be developed?

In apparent contradiction to previous speakers' contention that not enough Americans wanted to study computer or software engineering, U.S. enrollments in computer engineering had jumped to 78,000 in 2002 from 48,000 in 1998. But if a lot of people were again looking upon computer engineering as a good profession, would that continue into 2004 and 2005?

Analyzing the Optimistic Analysts' Optimism

Dr. Hira then called attention to a pair of studies that had been promoted by offshoring advocates and that he qualified as "optimistic." McKinsey Global Institute had claimed in a study called "Offshoring: Is It a Win-Win Game?" that the United States netted out 12-14 cents of benefit for each dollar spent offshore. Pointing to some of the study's limitations, he noted that no one had access to McKinsey's proprietary data and that the models used in the study were not explicit, so that there was no possibility of testing its assumptions using basic sensitivity analysis. Furthermore, its "very rosy" reemployment scenario—"an important part of its argument for how [offshoring] is a net benefit to the U.S."—did not hold up in the current labor market, nor did it account in the longer run for impacts on innovation and security that might be felt as change in the country's occupation mix exercised a chilling effect on both the software and engineering fields. Finally, McKinsey failed to disclose in its study several sources of potential conflict of interest: that it sells offshore consulting services; that India's software services trade association, NASSCOM, had been its client for some years; and that the former director of McKinsey was acting as the head of the U.S.-India Business Council. "I hope we're not resting our future on that particular study as being definitive," he remarked.

The second study, by Catherine Mann of the Institute for International Economics, based its optimism in part on the unrevised BLS occupation projection data (discussed above), as the updated figures had not yet become available when it came out. Dr. Hira called for reinterpreting this study in light of the more recent data; he also stated his disagreement with its contention that lower IT services costs provided the only explanation for either rising demand for IT products or the high demand for IT labor witnessed in the 1990s. He cited as contributing factors the technological paradigm shifts represented by the growth of the Internet, ERP, and Object-Oriented Programming; the move from mainframe to client-server architecture; and the upswing in activity surrounding the Y2K phenomenon.

Black-and-White Thinking Hinders Policy Debate

Turning to what he considered to be "policy-dialogue impediments," Dr. Hira argued that offshoring was affecting workers much more than it was companies, and that the latter felt no urgency to fix a problem they were not experiencing. Discussion of the issue, he said, should not be couched in the "good vs. bad" terms that had too readily defined it as a battle of "free trade vs. protectionism." Instead, the focus should be placed on assessing both the "good" and the "bad" that might result, and on how the potentially negative effects of offshoring might be mitigated.

Dr. Hira next displayed a table constructed using U.S. Department of Labor data from the labor condition applications (LCAs) that companies had to file in order to hire a foreign worker on an H-1B visa: the company's name, the title of the position it was seeking to fill, the place where the work would be done, and the prevailing wage for that position at that location (See Figure 40). The three companies listed above the bar were in fact engaged in onsite offshore outsourcing and offered annual salaries in the range of $21,000 to $33,000, which are significantly lower than the market rates for those occupations. Dr. Hira included Rockwell Scientific in the table because it had been put forward by industry advocates as a company that badly needed H-1B visas, which would enable it to hire exceptionally talented employees whom it was in fact paying $120,000 a

Company	Position	Location	Annual Wage
Accenture, LLC	Chief Programmer	Houston, TX	$25,113
Cognizant Technology Solutions	System Analyst	Tampa, FL	$32,870
Tata Consultancy Services	Programmer Analyst 1	Warsaw, IN	$21,460
Rockwell Scientific Company, LLC	Senior Scientist	Thousand Oaks, CA	$120,000
Rochester Institute of Technology	Assistant Prof of Economics	Rochester, NY	$60,000
Johns Hopkins University	Postdoctoral Fellow	Baltimore, MD	$30,500

FIGURE 40 Current H-1B and L-1 Visa laws enable and accelerate offshore outsourcing.
SOURCE: U.S. Department of Labor LCA database, <*www. flcdatacenter.com*>.

year to work at its facility in Thousand Oaks, Calif. In the interests of comparison, he also posted the positions of assistant professor of economics at his own institution, which pays $60,000, and of post-doctoral fellow in the biological sciences at Johns Hopkins, whose $30,000 salary level "might explain why a lot of people in America aren't too happy doing post-docs in life science." But it was the low level of the salaries being paid to workers from abroad that so many U.S. IT workers found problematic. "If work needs to be done in the U.S., it should be done by domestic workers unless you can't find a domestic worker," he stated, although he conceded that "maybe you can't find a domestic worker for $21,000." Also, the firms above the bar, all offshore outsourcing firms, are importing orders of magnitude more foreign workers on H-1Bs than firms like Rockwell Scientific.

Dr. Myers then introduced William Bonvillian, the Legislative Director and Chief Counsel to Senator Joseph Lieberman of Connecticut.

OFFSHORING POLICY OPTIONS

William B. Bonvillian
Office of Senator Joseph Lieberman

Mr. Bonvillian[36] began by posting what he called "two fighting quotes from the current times," one from Gregory Mankiw, the chairman of the President's Council of Economic Advisers (CEA), and the other from Intel Corporation Chairman Andy Grove (See Figure 41).

Mankiw, after loosing a "storm in Washington" with his pronouncement, had all but disowned it, but the points he had made initially were nonetheless classic defenses of outsourcing tendencies:

- that outsourcing was just a new way of doing international trade;
- that many things were more tradable than they had been in the past, which was good; and
- that outsourcing was on the rise and should be viewed as a plus for the economy in the long run.

[36]Mr. Bonvillian's views are further elaborated in his article, "Meeting the New Challenge to U.S. Economic Competitiveness," *Issues in Science and Technology,* XXI(1):75-82, 2004. See, also, Office of Senator Lieberman, White Paper: *Offshore Outsourcing and America's Competitive Edge: Losing Out in the High Technology R&D and Services Sectors,* May 11, 2004; Office of Senator Lieberman, White Paper: *National Security Aspects of the Global Migration of the U.S. Semiconductor Industry,* June 2003, pp. 1-10 (competitive pressure from China on a U.S. advanced manufacturing sector, semiconductors); Office of Senator Lieberman, *Data Dearth in Offshore Outsourcing: Policymaking Requires Facts,* December 2004 (data presented in his presentation was developed in part by his office for these three reports, which are carried on the Senator's Web site).

- "Outsourcing is just a new way of doing international trade. More things are tradeable than were tradeable in the past and that's a good thing.... I think outsourcing is a growing phenomenon, but it's something that we should realize is probably a plus for the economy in the long run." –Gregory Mankiw, CEA, 2/10/04

- "When you look at the software industry, the market share trend of US based companies is heading down and the market share of the leading foreign competitive countries is heading up. This X Curve mirrors the development and evolution of so many industries that it would be a miracle if it didn't happen in the same way with the IT service industry. That miracle may not be there." --Andy Grove, 10/9/03

FIGURE 41 Mankiw vs. Grove.

Grove's statement, which had been circulated widely in the nation's capital around the same time, reflected the other side of the coin. The two together provided a backdrop for the discussion.

Forrester had projected that 3.3 million U.S. IT service jobs and $136 billion in wages would go offshore over the following 15 years, while McKinsey had predicted a 30-40 percent annual acceleration over 5 years in the number of such jobs lost to outsourcing. Mr. Bonvillian said he had seen an even more extreme prediction: that a total of 14 million IT service jobs would disappear from the United States in that manner within a decade. He was skeptical of these projections because the relevant government agencies were not collecting the foundation job data. Nonetheless, while very significant declines in IT employment might be led by IT manufacturing, IT services would be right behind. Although the lack of data made it impossible to track the activity of the many companies engaging in overseas outsourcing, it was clear that the phenomenon was not restricted to any one sector. From low-end services—like call centers, help desks, data entry, accounting, telemarketing, and processing work on insurance claims, credit cards, and home loans—it was moving increasingly toward such higher-tech, higher-end services as software, chip design, consulting, engineering, architecture, statistical analysis, radiology, and health care centers. And those were only some of the leaders.

R&D Following Manufacturing Overseas

Another, parallel phenomenon, mentioned earlier by Dr. Hira, needed to be kept in mind: the growing trend of moving R&D offshore so that it would be closer to manufacturing (See Figure 42). The country had already lost 2.6 million manufacturing jobs, many of them in the tech sector during the 2001 recession and post-recession, and R&D had started to follow. A significant part of the R&D going abroad had been very high-end, very capable; in semiconductors, to pick one example, it was very important to have R&D and design close to the manufacturing stage. R&D spending abroad by U.S. corporations had quadrupled since 1968 to about $17 billion, and, since 1985, the ratio of foreign to domestic corporate R&D spending by U.S. firms had risen 50 percent. "These are big numbers," Mr. Bonvillian observed, which U.S. policymakers had to "begin to understand."

He then posted a pair of lists—of factors behind the exportation of U.S. jobs in services and R&D (See Figure 43), and of risks involved in transferring business functions abroad (See Figure 44)—without commenting on them other than to say that moving offshore is "not necessarily a simple equation [but] a complicated business transaction." Another graphic, this one illustrating the differences in salary levels for computer programmers from country to country (See Figure 45),

- Corporations are moving engineering, design, and R&D offshore to follow manufacturing.
- Corporate R&D funding accounts for 68% of all domestic R&D.
- Manufacturing sector is responsible for 62% of corporate R&D.
- Corporate R&D spending abroad quadrupled since 1986–now $17 billion.
- Ratio of foreign to domestic corporate R&D spending rose by 50% since 1985.

FIGURE 42 R&D trends.

- Cheap labor
- Large pool of educated & motivated workforce
- Instantaneous and low cost communications
- Foreign government investment in infrastructure
- Favorable business climate
- Improved protection of intellectual property
- Rising number of international collaborations
- Work around the clock (international time zones)
- Low cost computing, standardized software, digitization
- Access to large markets
- Established offshoring processes and models

FIGURE 43 WHY is the U.S. losing service and R&D jobs to overseas?

- Loss of in house expertise & future talent
- Hidden costs
- Political and financial stability of host nation
- Dependability of infrastructure in host nation (communication, energy, etc)
- Complex multicultural project management
- Shortage of English speakers
- Security and privacy
- Intellectual property protection

FIGURE 44 Risks associated with moving offshore.

was offered in explanation of the offshoring trend. Both it and a chart representing differences in the ratio of engineering degrees to total bachelor of science degrees in the United States and China, among other countries (See Figure 46), depicted what he judged "startling historical developments."

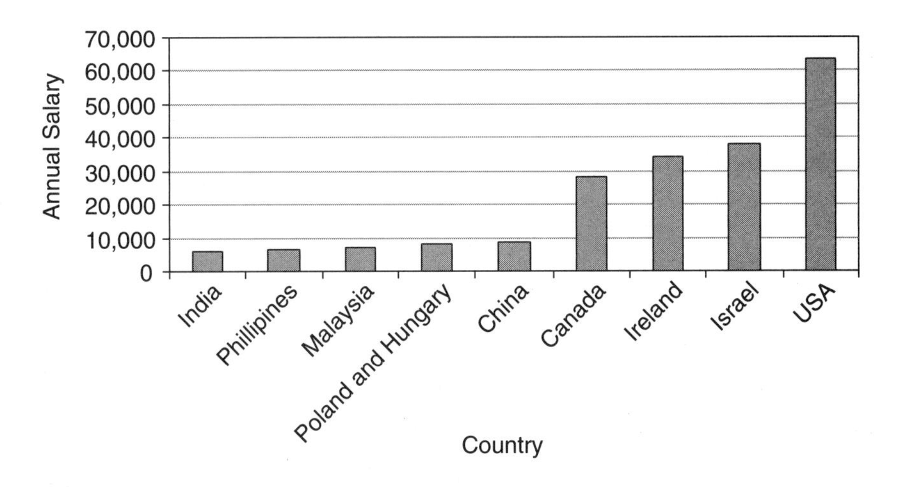

FIGURE 45 Annual salaries for a programmer in various countries.
SOURCE: *Computerworld*, April 28, 2003.

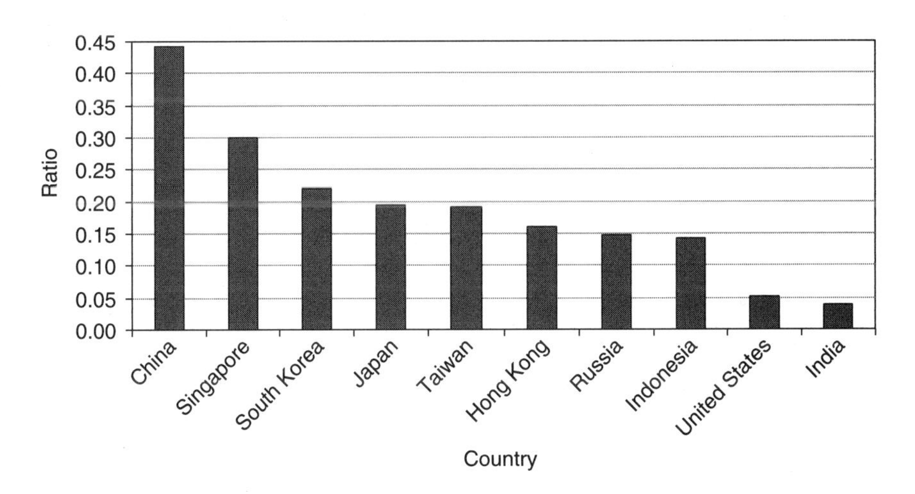

FIGURE 46 Ratio of engineering to total B.S. degrees awarded in various countries.
SOURCE: National Science Foundation, *Science and Engineering Indicators*.

While offshoring's impact on the U.S. economy was hard to ascertain using the limited data available, a debate had begun between optimists adhering to classical economics and some who had been voicing a more general but growing concern about the country's economic future. The former argued that offshoring would yield such benefits as lower product and service costs, new markets abroad fueled by improved local living standards, and more latitude for U.S. corporations to focus on core competencies at home. The other side, goaded by what Mr. Bonvillian dubbed the "Uh-oh Factor," and seeing growing world competition over innovation capability, cited downward pressure on wages for high-skill jobs, as well as diminishing talent, tax, and investment bases at home. Such concern, having become "in many ways a storm," was starting to buffet Capitol Hill.

Will Offshoring Erode Technological Comparative Advantage?

He then accorded special consideration to the argument that offshoring's movement up the value chain is accompanied by potential loss of technological competitive advantage. This could be seen as an emerging bone of contention between classical economics and the growth economics school. A representative of the latter, Clayton Christiansen of Harvard, had written that low-end entry and capability fuel the capacity to move into higher-end markets.[37] This familiar pattern of economic expansion, typified by the progression from the Corolla to the Lexus over time, had been replicated in any number of industries. "We have to understand that low-end competition now is not necessarily going to end at those low-end levels," Mr. Bonvillian warned.

Similarly, Michael Porter had contended that if high-productivity jobs are lost abroad, then long-term economic prosperity is compromised.[38] Some growth economists, echoing real estate agents, had placed the emphasis on "location, location, location." Porter himself had done a great deal of work on "clustering," the notion that it is possible to create a competitive force that is regionally based and collaborative across different sectors and institutions. The cluster spurs upgrading because it stimulates diversity in R&D approaches and provides a network mechanism for introducing new strategies and skills. Location is a key element: Since there is a tremendous skill set involved in the different work functions at an advanced-technology factory, losing a significant part of that manufacturing will affect the cluster's ability to thrive in its region. And when such loss is taking place on the service and R&D sides as well, another set of issues arises on top of that. According to John Zysman, who has maintained that manufacturing matters even in the Information Age, the advanced mechanisms for production and the accompanying jobs are a strategic asset, and their location makes the difference as to whether or not a country is "an attractive place . . . to create strategic [economic] advantage."[39]

---

[37]See Clayton Christianson, *The Innovator's Dilemma*, New York: Harper Business, 2003.

[38]Michael Porter has argued that business leaders should look for locations that gather industry-specific resources together in one "cluster" that can lead to competitive advantage. See Michael Porter, "Building the Microeconomic Foundations of Prosperity: Findings from the Business Competitiveness Index," *The Global Competitiveness Report 2003-2004*, X Sala-i-Martin, ed., New York: Oxford University Press, 2004.

[39]See Stephen S. Cohen and John Zysman, *Manufacturing Matters: The Myth of the Post-Industrial Economy*, New York: Basic Books, 1988.

"A Different Kind of Competitiveness Story"

What all this added up to, Mr. Bonvillian said, was "a different kind of competitiveness story." Although the debate outlined above might seem familiar from the 1980s, a new set of competitors had emerged and the competitive situation had grown far more complicated:

- In its competition against Japan during the 1970s and 1980s, the United States was facing a high-value, high-wage advanced-technology country very similar to itself, whereas competing against China meant competing against a low-wage, low-value increasingly advanced-technology country.
- The United States, one could argue, had been able to use its entrepreneurial advantage to offset Japan's advantage in industrial policy. China, in contrast, was not only a very entrepreneurial country, it was making intense use of industrial policy in pursuing such aims as capturing the semiconductor sector.
- The rule of law, which was a common assumption in competition with Japan, is still an emerging idea in the competition with China. By a similar token, Japan protected intellectual property, while an "intellectual property theft model" unfortunately structured much of the Chinese competitive environment; according to one source, the FBI estimated at around $250 billion per year the intellectual property theft in China of U.S.-bred products and ideas
- Japan was a national-security ally, China a potential peer-competitor.

One constant had been that both Japan and China had undervalued their currency and bought U.S. debt, in consequence of which the United States was not in a position to push its trade arguments vigorously. China would maintain as long as it could the low value of its currency while continuing to buy U.S. debt at massive levels, he predicted, so it can retain leverage over U.S. economic policy.

With data scarce and concern "enormous," a multitude of bills had been introduced in Congress, which, according to Mr. Bonvillian, sometimes reflected a move toward a protectionist outlook. Under a bill offered by Sens. Edward Kennedy and Tom Daschle, companies that sent jobs abroad would have to report how many, where, and why, giving 90 days' notice to employees, state social-service agencies, and the U.S. Labor Department. Senator John Kerry had introduced legislation requiring call-center workers to identify the country they were phoning from. Other bills would require government contractors to have 50 percent of their employees in the United States, prohibit work under federal contracts from being performed outside the country, or bar companies that outsourced jobs from contracting with the federal government. Legislation that had failed three years before, but was being revived would extend Trade Adjustment Assistance (TAA), up to then available only in the case of manufacturing jobs that had been offshored, to service-sector jobs as well—a proposition bringing with it a complex definitional issue.[40] And the Genuine U.S. flag Act, one of his personal favorites, would prohibit the purchase of American flags made outside the country.

Discovering That Services Are Not Invulnerable

U.S. policy makers' previous exposure to manufacturing issues had helped build their sophistication in responding to competitiveness problems in that sector. But, not having envisioned significant competitive threats to the country's service sector, which had been regarded as "golden," and invulnerable, they found themselves confronting what Mr. Bonvillian called a "complicated dilemma." After taking the initial step of collecting data, lawmakers would be obliged to "think about some safety nets" in light of the fact that the voices of their constituents had grown "loud" on the subject. Among the areas of potential response in the near term were:

- **Retraining.** "A much more effective, rapid, and agile-workforce training set of programs," with interactive IT playing a role, would be required.
- **Compensation.** Options included Trade Adjustment Assistance (TAA) for services; job-loss insurance, which had been featured in a pilot program in previous TAA legislation; and asset building compensation features. Novel thinking about incomes and assets was in order, Mr. Bonvillian said, because "in the kind of economy we've got now, holding assets is probably much more enduring than simply relying on straight incomes. Our upper-middle class has that capability, but it doesn't go very far down the chain. And how do we start to turn that perspective around in governmental policy?"
- **Trade.** The large U.S. apparatus for negotiating trade deals had limited capability for looking at ongoing and shifting barriers to market entry that U.S. companies faced, and for prompt action against unfair competitive practices. Much hard work was ahead, starting with attempting to cope with "a tremendous amount of just straight industrial policy . . . that is likely GATT violative," such as the VAT rebate China was providing on domestically manufactured semiconductors.
- **Financing.** Means needed to be found for bringing spending from the venture-capital system off the sidelines and back into the economy.

In the longer term, if economic growth theory was right, the country would have no choice but to innovate its way out of the situation. As a consequence, far

[40]Trade Adjustment Assistance for Firms (TAA), a federal program, provides financial assistance to *manufacturers* affected by import competition. Sponsored by the U.S. Department of Commerce, this cost sharing federal assistance program pays for half the cost of consultants or industry-specific experts for *projects* that improve a manufacturer's competitiveness.

more serious thought had to be given to the national innovation system, raising such questions as:

- How might the United States increase the speed with which it brings on the "next big things"?
- How should it cope with talent issues surrounding math and science education?
- How could the number of "prospectors" in the nation's innovation system be increased, and how could they be given the skill sets not just to make discoveries, but to grow companies?
- Could the country invest in R&D in such a way as to spur the development of targeted new technologies, especially in the physical sciences? If the U.S. is going to have to compete in high-end services, could it build up its negligible services R&D and accelerate services innovation?
- What could be done to break down barriers to entry for truly high-speed broadband which could spawn a new generation of IT applications, and how should growth of this broadband infrastructure be augmented?

DISCUSSION

Charles Wessner of the STEP Board pointed to a disjunction between the ability of both the economy and the policy process to adjust and the time required to take any of the long-term steps listed by Mr. Bonvillian. The lag effects of 7 years of substantial cutbacks in the federal R&D budget, which STEP had documented, were expected to hit.[41] A suggestion had been made that even were the United States to pursue a trade case against China, around 18 months would be required to file it—the "equivalent of letting people rob the bank for 18 months before you pick them up"—and China would have completed planned investments on its 300 mm wafer-fabricating plants by then. "We know there's a fire, and we're not able to bring hoses to bear," he contended, characterizing U.S. policy as "bankrupt." At the same time, although Europe appeared so much less agile than the United States, European nations boasted very high standards of living and employment rates with high-wage, high-welfare jobs. Although Europe did have a structural unemployment problem, with the rate somewhat higher than that in the United States, European countries were in a position to pay their residents not to work. "Do they have any secrets?" he asked. "How do they do it?"

Mr. Bonvillian responded that Europe's safety-net structure was much more profound than that of the United States, whose government was so limited by long-term fiscal obligations that it did not have the resources to construct a new

[41]National Research Council, *Trends in Federal Support of Research and Graduate Education*, Stephen A. Merrill, ed., Washington, D.C.: National Academies Press, 2001.

set of safety nets. He said, however: "If you have to choose between some key investments in safety nets vs. some key investments in your innovation system, I'd put the money on the innovation system." He acknowledged the existence of serious short-term concerns and conceded that the economic recovery then taking place was of a very different kind from any he'd seen before: Companies were recovering while jobs were not, a phenomenon that might turn out more enduring than many suspected. But if growth economics proved right, and the bulk of growth came out of technological and technologically related innovation, then the only course of action was to get the nation's "innovation house" in order.

Can the United States Innovate Its Way Out?

Egils Milbergs of the Center for Accelerating Innovation observed that the issue of employment was creating political pressure for President Bush that was compounded by the statistical difficulty of making job forecasts. There had been a loss of jobs owing to offshoring, and the BLS's job-market projections, in particular those for IT jobs, had kept coming down. Automation, much of it driven by software, was causing productivity to increase and, at the same time, shrinking the need for workers. According to a recent study by the New York Federal Reserve Board, 75 percent of layoffs were now permanent—up from around 50 percent in previous downturns—so jobs would not be restored with a cyclical upturn. In short, the job outlook was "pretty downbeat." Referring to the prospect mentioned by Mr. Bonvillian of "innovating out of the problem," Mr. Milbergs suggested that Wal-Mart employed more people than had the entire Internet economy, even if the latter could be credited with generating 1 million jobs. "I'm wondering how we see ourselves out of this," he stated, asking whether the situation on the employment front could be expected to turn around in the following 18-24 months.

In response, Dr. Hira noted that recent employment data demonstrated not job creation but job destruction, and he speculated that, if the reverse were true, offshoring might be less controversial. Not unlike economists, he was unable to pinpoint the causes of the trend; therefore, he was hesitant to try projecting even 18 months out. Rather, he saw the question thus: "What do you do now, and how much damage does this current situation create for future innovation?" Evidence of the climate's chilling effect on engineers and software developers, he said, was that they tended to converse among themselves about offshoring instead of about attending technical conferences and advancing technology, and they did not feel their concerns had met with a straightforward response. "They've been told over and over again, 'This is actually good for you, because it frees you up.' Well, it frees you up to do what?"

Dr. Varian had two points to contribute to this discussion:

- **"Be wary of focusing on just one industry."** There were different trends in different sectors, he said, citing a recent report in *The Economist* that the majority of biotech research was moving to the United States because world-high prices for pharmaceuticals made it attractive to sell here and because of the importance of locating production close to the market.

- **"Look at the medium term."** Focusing on 7 years out, rather than on 2 years out or on 10-15 years out, would bring into view the huge labor-market upturn that would be hitting with the baby-boomers' retirement. That would create job loss—"quote-unquote"—for voluntary rather than involuntary reasons, and the country would need skilled labor. While he recognized that those currently unemployed might find little solace in this prospect, he said that the kind of policy responses previously mentioned—encouraging education in America of engineers, loosening some restrictions on technological development—would be very important in 7, 8, or 9 years.

Concluding the panel, Dr. Myers said that a speech by Alan Greenspan then posted on the Federal Reserve Web site would probably be worth keeping in mind while seeking solutions. He paraphrased the Fed chairman as saying that the nation had learned over the previous 50 years that a flexible and adaptive economy was the most robust with respect to unexpected events leading to downturn. Dr. Myers interpreted this to mean that the solutions base had to be adaptive rather than fixed in such a way that the economy could not adjust to the changes that were bound to occur. He then thanked the speakers for their presentations on what he deemed a very provocative subject.

Participants' Roundtable—Where Do We Go from Here? Policy Issues?

Moderator:
William J. Raduchel

Dr. Raduchel opened the roundtable by voicing his hope that the attendees would, above all else, take away from the conference an understanding of the uniqueness and complexity of software, as well as of the ecosystem that builds, maintains, and manages it. It is because one does not "really know what's happening through that whole ecosystem that brings down the software," he said, that an issue like offshoring "gets so complicated so quickly." The day's second most important point was that the way software was created, maintained, and distributed had not changed in 40 to 50 years. It should come as no surprise, therefore, that other countries had learned the prevailing techniques and become competitors of the United States. Unless there were investment that fundamentally changed the ways in which software was created, the industry would have to be considered mature even though what it produced was at the cutting edge of advanced technology.

Dr. Raduchel then asked each member of the panel to take two minutes to identify the one issue relating to software that he would put at the top of a public-policy agenda for the United States. He called upon Wayne Rosing of Google to go first.

Wayne Rosing
Google

Dr. Rosing related that he had posed the following question to four or five friends of around his age who had lost jobs in the dot-com bust: "When was the last time you taught yourself something new—that you really took the trouble to learn a new field?" The initial response, he said, was most frequently "a blank stare," followed moments later by an answer of 10, 12, or 15 years ago. That many individuals did not stay current was one of the biggest single problems the United States faced, because in a global economy those who do not keep their skills at the cutting edge are inevitably sidelined. What the country needed was a system that created extraordinary incentives for people to take charge of their own careers and their own marketability. Much change would be necessary because there was a lot that did not work well; the colleges, for example, were designed for young, full-time students. Awareness would have to be built, but the situation was not without opportunity.

Kenneth Flamm
University of Texas at Austin

Noting that the day had begun with "lofty questions about 'What is software?' " and concluded with "a somewhat tense discussion" of offshoring, Dr. Flamm proposed to offer four points tracing the route from one to the other. The first harked back to the history of the semiconductor industry, which became one of the United States' first to go offshore when, in the mid-1960s, virtually every manufacturer moved the unskilled, labor-intensive assembly part of the production process to Hong Kong. A decade later the only assembly that remained in the United States involved relatively short runs of very specialized product that had to be fabricated close to the market. Meanwhile, the commodity production had stayed in Hong Kong only until wages for unskilled labor were bid up there, when it moved to Korea. The same pattern was repeated several times: Production moved on to Taiwan, then to the Philippines, then to Malaysia, each time staying only until wages had gone up. Even in an area in which one could expect a relatively undifferentiated labor input going into the production process, the supply of unskilled labor was not infinitely elastic.

Second, Dr. Flamm maintained that the offshoring of software services in many respects greatly resembled the semiconductor story of the 1960s, the exception being that very skilled labor rather than unskilled labor was now involved. Central to the movement of semiconductor manufacturing abroad had been the growth of the air-traffic and telecommunications infrastructures. Transport costs are very low for the semiconductor because of its small physical size and light weight; and, by the mid-1960s, telecom had developed to the point where it was possible to coordinate production abroad and to do so in a cost-effective way. "Similarly," he observed, "if ever you are going to find a service that can be offshored economically, it's going to be software, which is not really a physical product but just bits flowing through wires."

WHY THE ABRUPT EMERGENCE OF THE OFFSHORING PHENOMENON?

Third, and with this in mind, Dr. Flamm asked why offshoring seemed to have suddenly swept the software industry in the previous 2 or 3 years. Why hadn't this happened 5 years before? He speculated as to two potential answers. The first was that major investment in communications infrastructure outside the United States might have equalized the playing field, helping make this possible for the first time. The second was that the skill of the U.S. labor force had declined relative to that of its competition. He offered his impression as a member of a university faculty that the nation had not truly been investing in its educational infrastructure, and he asserted that the California public school system of the 2000s was a shadow of that "glorious and wonderful" system that had "invested all kinds of resources" in him when he went through it in the 1960s. He suggested that the diminution of resources available to public education in California had been replicated across the country, and that it had had an impact on the relative attractiveness of hiring U.S. workers and their competitors in other countries.

Fourth, the semiconductor industry had become subject to another round of competition in the 1980s, this time from other advanced industrial economies in a process very similar to that affecting the contemporary software industry. In this second round relatively skilled jobs rather than unskilled jobs had been at stake. The U.S. semiconductor industry had ended up coming back, and while part of the reason was that it had invested in resources, the bottom line was that it never returned to the commodity products it had gotten out of. It had instead gone into design-intensive products and stayed ahead.

Dr. Flamm drew the following policy prescriptions from this story:

1. **"Don't expect to get back into what you've been pushed out of."** The rest of the world has competence, and those nostalgic for the old monopoly should remember that the rest of the world had caught up and refrain from looking back.
2. **"Invest in the things that our new competitors invest in,"** which make economic development more accessible for all to participate in. While he was skeptical of the widespread notion that an infinitely elastic supply of top-notch university graduates was coming out of the schools of China and India, he argued that if the United States really wanted to compete, it had to invest in education. Likewise, to make less-industrialized areas of the country more attractive and to keep workers from the departing industrial sector employed, domestic investment in communications infrastructure would be necessary.

Ernst R. Berndt
MIT Sloan School of Management

Seconding Dr. Varian's earlier point that not all advanced technological sectors were behaving alike, Dr. Berndt observed that while enrollment in MIT's electrical engineering and computer science departments was indeed down, that was not true of fields related to biotechnology, such as biology and chemistry. At 30 years old the PC industry was, after all, no longer sexy but mature, and the action seemed to be elsewhere. Novartis was moving 900 scientists from Switzerland to Cambridge, Massachusetts, which he saw as part of "a Schumpeterian process that we just have to recognize and try to exploit." One of the best investments in this regard would be in getting better data.

James Socas
Senate Committee on Banking

Congressional staffers like himself, Mr. Socas stated, had been hearing every day from people who were losing their jobs and who could not afford the luxury of retraining themselves, as they were working two jobs or caring for small children or a sick parent. Such people reported seeing U.S. companies send offshore training and capital, resources that "used to stand behind U.S. workers"; as a result, workers abroad were being armed with all the skills that formerly had allowed U.S. workers to command the world's highest wages. "U.S. workers were paid more not out of the kindness of the heart of companies," he declared, "but because they produced more"—an ability now being transferred to their counterparts in other countries.

The "relentless race for profits" might be forcing U.S. companies into this, Mr. Socas conceded. But he recalled the phrase "what's good for General Motors is good for the country," which he interpreted as a "bold—and at the time arrogant but valid—statement" of GM's intention to invest capital and technology in the community of Detroit and its workers "to help them be the best that they could." In return, the U.S. government gave General Motors tax benefits for R&D and other advantages, such as the benefits of a strong public education system, and as a consequence both General Motors and the United States prospered. The substance of the complaint that the American workers had been bringing to Capitol Hill was: "It seems like this whole deal has been broken." Since they understand how the world works, being familiar with similar stories from agriculture and manufacturing, they were not so much objecting to the loss of a specific job as to what they saw as evidence that "the game has changed."

FREE TRADE AND PROTECTIONISM IN U.S. HISTORY

Such economists as Gregory Mankiw, chairman of the President's Council of Economic Advisers, had reached the conclusion that this change was a good thing, Mr. Socas said. Americans have been raised to believe that free trade is good for the economy in that it will cure displacement and many other ills. But because the country had endorsed free-trade policies over the previous half-century, an important fact was being left out of account: that its history is in fact one of protectionism, that McKinley ran on a protectionist platform, that the Civil War was fought over protection and tariffs.

Free trade is a wonderful policy nevertheless. Offshoring, however, is not free trade, and the current landscape was very different from the one that David Ricardo envisioned when he put forward his Theory of Comparative Advantage. In a nutshell, this theory states that each country should use its internal cost ratios and direct its productive resources to what it does most efficiently. This does not mean that a country needs to produce something at the lowest cost around the world; simply, a country must do something well and direct its resources thereto. The consequence will be that, through trade, world output will increase and, with it, the economic pie.

SEEKING COMPARATIVE ADVANTAGE OR ABSOLUTE ADVANTAGE?

But there is an assumption in Ricardo that, Mr. Socas claimed, nobody ever talks about: that the factors of production remain immobile. If instead they can leave a country, then they will not go from South Carolina to California but from South Carolina to Guangzhou Province or from South Carolina to Bangalore. They will no longer give the United States comparative advantage but will chase what Ricardo called "absolute advantage." In his famous example of wine and wool trading between Portugal and England, Ricardo lays this out very clearly. But Ricardo notes that the "factors of production" would never leave England, for no English capitalist would ever invest so substantially in another country out of loyalty to his mother country and out of fear for the security of his investments overseas.

Expressing skepticism that anyone in attendance believed this economic patriotism still applies today, Mr. Socas declared: "It certainly didn't apply in my life when I was in the world of investment banking, it certainly doesn't seem to be applying with those who talk about corporate strategies today." Even though that fundamental assumption was no longer valid, advocates of free trade continued to fall back on Ricardo to defend their position. This raised two questions: Was off-shoring something old, meaning the latest chapter in the history of free trade, or something brand new, the first manifestation of a globally integrated economy? And, if the latter were true, were the nation's policies—reflected in everything from the way the Bureau of Labor Statistics collected data to the way corporate accounting was handled—pinned to a vision of trade oriented toward a national system, when in fact companies were oriented towards a global economy? He called the latter "the central political question of our time."

WHAT DOES "PROTECTIONISM" AIM TO PROTECT?

Mr. Socas reiterated that he had observed a high level of grass-roots concern about these issues, adding that "average people" believed politicians to be exploiting them. The negative epithet "protectionist" obscured the question of what was being protected—the answer to which, he asserted, was the American system and the American community. He rejected a comparison between current reservations about free trade and the motivations for the "Buy American" campaigns of two decades before, characterizing the latter as efforts to shield Ford from its failure to invest in just-in-time manufacturing or to reward "workers that [had] lost their way." Now at issue was giving U.S. workers, who had suddenly been exposed to many new competitors worldwide, a shot at trying to compete. They needed the skills that would allow them to continue to command the high wages and opportunities "that, frankly, the whole structure of our country is founded on."

Ronil Hira
Rochester Institute of Technology

Declaring himself free of the orthodoxy shaping debates surrounding the issue of "free trade vs. protectionism," Dr. Hira observed that both terms were somewhat vague and that they tended to be "situationally implemented." As to the specific issue of software, he recalled the 1980s question of whether manufacturing matters, asking whether software matters. "If it does," he stated, "we need to think about how to go about making software a viable profession and career for people in America," since the field did not appear sufficiently attractive at present. He expressed his impatience with what he termed "the old stories of 'We just have a bad K-12 math and science education' " as the reaction to a call for policy alternatives. While educational improvement was in order, it was also important to think through and debate all possible options rather than tarring some with a "protectionist" or other unacceptable label and "squelching them before they come up for discussion."

Concluding Remarks

Dale W. Jorgenson
Harvard University

Expressing his appreciation to all involved, Dr. Jorgenson rated the day's session as far exceeding expectations that had been high to begin with. He extended special thanks to Dr. Raduchel, whose intellectual leadership on behalf of the STEP Board had been reflected in the meeting's high quality; and to Dr. Wessner, as well as to his staff, for putting together a stimulating program of discussion on very interesting issues.

Dr. Jorgenson noted that the subject of the series of which this conference was a part, "Measuring and Sustaining the New Economy" was an area discovered by economists only in 1999, when, the U.S. Bureau of Economic Analysis (BEA) had introduced the capitalization of software. This was "really quite remarkable" in light of the fact that the computer had entered into commercialization some four decades before. Since economists had been lagging in that area, a matter on which various viewpoints had been provided by the day's group of distinguished speakers, the question confronting them was: "How are we going to fill this gap?"

UNDERSTANDING THE ECONOMICS OF THE COMPUTER SECTOR

The first necessity was understanding what the subject was and what role it played, issues that had been discussed very elegantly by Dr. Raduchel. Dr. Lam had then imparted a great deal of knowledge about the current frontier of computer science. "We have to understand as economists," commented Dr. Jorgenson, admitting that this was a "parochial" view, "what computer science is and what computer engineering is, and what the difference is, and how that is going to develop." Economists like Dr. Varian were taking the lead in trying to understand the economics of this subject, which was both "very, very important" and, for economists, "very, very new."

Quite a bit had been accomplished in the course of the day, as a "very firm story" had been laid out concerning the unbelievably rapid progress being made in the science and technology of the area. Drs. Berndt and White had provided guidance through the landscape of prepackaged software, discussing where to measure it: at the Microsoft gate as it leaves Redmond, Washington, or when it arrives where somebody is actually going to put it to use. Taking quality change into account was extremely important, Dr. Jorgenson observed, even if that might be objected to as a "pretty esoteric point." While conceding that it was an example of economics jargon, he noted that taking quality change into account involved incorporating computer science and computer engineering into the economics of software. "That is our agenda," he declared, praising Dr. Berndt and White for their elegant illustration of it.

ACCOUNTING FOR COMPUTER-RELATED COSTS: A "GOOD-NEWS STORY"

Dr. Jorgenson then addressed the concern lest this enterprise be "just too complicated." It may start out with a mere shrink-wrapped package, but how could measurements capture such associated costs as installation, business reorganization, and process reengineering? "Calmer heads prevailed," he assured the audience, and the result had been "a thoroughly good-news story." He recalled the discussion of accounting rules by Mr. Beams and Ms. Luisi, which had made clear that these problems had been thought through and that agreement had been reached on accounting rules and on how they were to be applied. "Admittedly, there are a lot of ambiguities," he said, "but the accountants are the people we depend on, and they have delivered."

There was more good news. As Mr. Wasshausen had mentioned, the U.S. Census Bureau, "in its wisdom" and "just in time," was fielding a first-ever survey that would determine where investment in software was going, how much software was being produced in the United States, how much was being imported, and how much the country was exporting. The results of the survey were to become available in the first quarter of 2004. "Wait a minute!" Dr. Jorgenson exclaimed. "We discovered this problem in 1999, and only 5 years later we're getting the data!" Furthermore, that data were to be certified by the Financial Accounting Standards Board.

BUILDING UNIFORM DATA INTO NATIONAL ACCOUNTS

Dr. Pilat, he noted, had "started off on a somewhat sour note" by describing the picture across the Organisation for Economic Co-operation and Development (OECD), whose member countries have different methods of accounting, as "pretty chaotic." But an OECD task force has delivered a report, and all the national statisticians have gone back to their home countries to mount surveys in the aim of beginning to build these data into their national accounts in the way that the BEA had built them into the U.S. national accounts beginning in 1999. This meant that international comparisons were in the offing, even if they were not to be expected right away. Their availability—that Dr. Jorgenson foresaw within 12 to 24 months—would make it possible "to supply the missing link: moving offshore." It would then be possible to ascertain what was moving where. There had been unanimous agreement among the members of Panel IV that the starting point for any discussion had to be that data were not yet available. But the data were on the way, which meant that policy would not have to be debated without the illumination of careful economic measurement.

Dr. Jorgenson again thanked all participants for their contributions to what he called a "very clearly focused picture of the challenges that lie ahead of us, the opportunities, and the potential resolution of what has become a very, very tense and therefore a very interesting debate."

III

RESEARCH PAPER

The Economics of Software: Technology, Processes, and Policy Issues

William J. Raduchel

Software is the medium through which information technology (IT) expresses itself on the economy. Software harnesses and directs the power of hardware, enabling modern data analysis and a variety of application domains. While software is becoming increasingly central to our productive lives, it is also becoming increasingly hard to create, increasingly hard to maintain, and increasingly hard to understand.

Software represents the core of most modern organizations, most products and most services. The U.S. military believes that the majority of cost in future weapons systems will be software. Software embodies the knowledge and practice by which the organization performs it mission and consists of a design and an implementation.

Software today may well be the largest single class of assets in the world, but it is intermingled with many other components, so distinguishing its contribution is not simple. So, for reasons discussed below, it is not measured that way. It is also undoubtedly the largest single class of liabilities. Operating, maintaining and creating software is surely the largest single class of expenses other than direct labor. Enterprise Resource Planning systems determine the way that factories run. Software and the business practices it enables are the biggest single driver of productivity growth.[1] According to Dale Jorgenson, IT accounted for more than half of the one percentage point increase in the Gross Domestic Product (GDP) growth rate between 1990-1995 and 1995-1999.[2] And, as we learn more every week, software by far creates the greatest vulnerabilities to our economy.[3] The primary driver of the August 14, 2003, northeastern U.S. blackout, which cut off electricity to 50 million people in eight states and Canada, was a deeply embedded software flaw in a General Electric energy management system.[4] The U.S. economy is so dependent on software in ways that we currently do not understand.

Software drove the great bubble of the 1990s, which created more wealth than anything else in history.[5] Software created millions of new high-paying

---

[1]"The software sector is one of the most rapidly growing sectors in OECD countries, with a relatively strong performance across all economic variables. The sector contributes directly to economic performance because of its dynamism, and software applications help boost growth across the whole economy through their use in an ever-expanding array of applications. Rapid growth in the sector is evident in terms of value added, employment, wages, R&D intensity, patents and investment." Organisation for Economic Co-operation and Development, *Information Technology Outlook 2002: The Software Sector*—Growth Trends, Paris: Organisation for Economic Co-operation and Development, 2002, pg. 105.

[2]Dale W. Jorgenson, "The Promise of Growth in the Information Age," The Conference Board, Annual Essay 2002, pg. 4. "For most OECD countries . . . software was the major source of increased investment in knowledge during the past decade." Organisation for Economic Co-operation and Development, *Strengthening the Knowledge-based Economy*, Paris: Organisation for Economic Co-operation and Development, 2002, Chap. 1, pg. 25. Oliner and Sichel showed that two-thirds of the increase in productivity between 1990-1995 and 1995-1999 is attributable to IT. S. D. Oliner and D.E. Sichel, "The Resurgence of Growth in the Late 1990s: Is Information Technology the Story?" *Journal of Economic Perspectives*, 14(4), 2000. "For most OECD countries . . . software was the major source of increased investment in knowledge during the past decade." Organisation for Economic Co-operation and Development, *Strengthening the Knowledge-based Economy*, 2002, *op. cit.* "[G]rowth accounting estimates show that ICT [information and communications technology] investment typically accounted for between 0.3 and 0.8 percentage points of growth in GDP per capita over the 1995-2001 period. . . . Software accounted for up to a third of the overall contribution of ICT investment to GDP growth in OECD countries." Organisation for Economic Co-operation and Development, *ICT and Economic Growth: Evidence from OECD Countries, Industries and Firms*, Paris: Organisation for Economic Co-operation and Development 2003, pg. 36.

[3]"Identified computer security vulnerabilities—faults in software and hardware that could permit unauthorized network access or allow an attacker to cause network damage—increased significantly from 2000 to 2002, with the number of vulnerabilities going from 1,090 to 4,129." *The National Strategy to Secure Cyberspace: Cyberspace Threats and Vulnerabilities*, Washington, D.C.: Executive Office of the President, February 2003, pg. 8.

[4]Kevin Poulsen, "Software Bug Contributed to Blackout," *Security Focus*, February 11, 2004. Another example is the pair of software flaws that caused total mission failure for two consecutive NASA Mars missions.

[5]In the knowledge-based economy, "the creation of wealth becomes synonymous with creating products and services with large software content." J. Hagel and A. G. Armstrong, *Net Gain,* Cambridge, MA: Harvard Business School Press, 1997.

positions[6] and redefined the way hundreds of millions of others do their jobs. Over the next decade, eight of the ten fastest growing occupations will be computer related, with software engineers comprising the fastest growing group.[7] This prediction may be at risk, as we see software emerging as a key way India and China are upgrading their economies by outsourcing these jobs from the United States, where computer science continues to decline in popularity as a field of study.[8] Immigration policy finds it hard to distinguish in favor of the truly talented developers, so firms now outsource whole projects instead of sponsoring a few immigrants.

There is currently no good economic model for software and, consequently, we have no good understanding of software. This paper is an attempt to start a search for what is a good economic model for software. This is an attempt to state the problem, in the belief that all good work begins with a correct problem statement. It does not attempt to answer the questions raised. It does distill over 40 years of personal experience creating, maintaining and using software and managing others doing these tasks from multiple perspectives over a varied career.

THE FUTURE

The future will only increase the importance of software to our society and economy. The costs of rendering, processing, storing, and transmitting digital information—whether data or pictures or voice or anything—continues to decline in all dimensions and likely will continue to do so for decades.[9] We are not at the end of the computer revolution at all. No industry is safe from reengineering as cheaper computing enables ever more sophisticated software to redefine business practices.

A firm is an information system—one composed of highly decentralized computing by fallible agents called people. In many cases, the value of a firm is entirely tied up in the value of its software systems. Software defines the informa-

---

[6]In the United States alone, more than 1.2 million jobs were created in the software sector from 1990-2001. Furthermore, "jobs in the software sector alone are only part of the contribution of software activities to total employment. In the United States, all employment in computer and related services accounted for only 1 percent of total employment in 1998 (including the public sector), whereas the share of computer-related occupations in the economy was around 2 percent. As a result, as much as 76 percent of computer-related occupations were in other industries." Organisation for Economic Co-operation and Development, *OECD Information Technology Outlook 2002: The Software Sector*—Employment, Paris: Organisation for Economic Co-operation and Development, pg. 106-107.

[7]"Fastest growing occupations, 2000-2010," Office of Occupational Statistics and Employment Projections, U.S. Bureau of Labor Statistics, accessed at <*http://www.bls.gov/news.release/ecopro.t06.htm*>.

[8]Michelle Kessler, "Computer Majors Down Amid Tech Bust," *USA Today*, October 8, 2002.

[9]Trends recently reviewed in the STEP Workshop on Deconstructing the Computer. See National Research Council, *Deconstructing the Computer*, Dale W. Jorgenson and Charles W. Wessner, eds., Washington, D.C.: National Academies Press, 2005.

tion system, and increasingly the value of a firm is embedded in its brands, its contracts and its systems, all intangible. Tangible assets increasingly are rented on an as-needed basis because software enables this. However, these software systems, the most key assets for companies, are not reported on balance sheets because to the accounting profession software is intangible and hard to measure. The result is that investors are left in the dark about a company's single most valuable asset. I for one fail to understand how that is in the best interest of investors.

CONSUMERS

The growing importance of software can be seen in the way that it is redefining the way consumers communicate, entertain, and inform themselves. Voice telephony, music, and movies are all rapidly becoming just another application. The number one device for managing and playing music in the world is the personal computer, followed by devices like iPods that connect to it. The very definition of music has been changed in the past 5 years, and the music industry has to totally reinvent its business model as a result.[10] Personal computers, mobile telephones, and consumer entertainment devices are morphing together in ways that will wrench the communications, entertainment, and publishing industries for decades to come. The size of the gaming industry, valued around $35 billion, is rapidly approaching the $38 billion music industry and has already surpassed the motion picture industry and continues to grow its hold on the leisure time worldwide.[11] The growth of gaming is so large that eventually we will rewrite almost all software that interacts with consumers into a gaming paradigm.

Piracy is going to remain a recurring issue because the entertainment industry makes a profit by delivering content to consumers in ways that consumers do not want it. The reason that piracy exists is because consumers want music organized by a mix of tracks on a single playlist. The music industry has refused to provide music in this way because it does not suit any of their business models.

NATURE OF SOFTWARE

Software, on the one hand, looks remarkably easy and straightforward, and to some extent it is. The challenge is making software that is failure-free, hazard-free, robust against change, and capable of scaling reliably to very high volumes, while integrating seamlessly and reliably with many other software systems in real time. I recently was the Castle Lecturer on Computer Science at West Point

---

[10]Kevin Maney, "Music Industry Doesn't Know What Else To Do As It Lashes Out at File-sharing," *USA Today*, September 9, 2003.

[11]Loren Shuster, "Global Gaming Industry Now a Whopping $35 Billion Market," *Compiler*, July 2002.

Academy and many of the cadets were pretty smug at finishing their first programming exercise of 50 to 100 lines. That example is writ large, as the problem with software is not creating a piece of code, but rather, creating one that will work well over time. The robustness of software is not only an issue of sloppy coding. Many software flaws are deeply rooted and substantive vulnerabilities can come from misunderstandings about the true requirements for a system, its environment of use, and other factors. The errors allowing most recent security breaches are relatively simple and easily eradicated.[12]

Real-world software is hundreds of millions of lines of code that make up the applications that run big companies. Those applications are resting on middleware and operating systems that, in turn, are tens of millions of lines of code. The complexity is astounding. The software stack on a modern personal computer is probably the most complex good ever created by man.[13] The average corporate IT system is far more complicated than the Space Shuttles or the Apollo project, but engineered with none of the rigor or disciplines or cost of those programs. Bill Joy calls the spiraling complexity putting "Star Wars-scale" software on the desktop.[14]

The costs of maintaining and modifying software increase over time and increase with the amount of accumulated change because as modifications begin, increasing complexity is introduced and eventually, changes cannot be made at all. The biggest issues we face lie in how we administer and update our software, and we have a situation where software is sold without liability and where the markets are heavily dominated by one or two vendors. Competition is unlikely to work its usual magic to solve this problem.

Software has a useful life. Because the underlying technology changes, eventually the design needs to change as well. Modifications to the software make it harder to further modify. Seven years is a good starting point for software life, but few organizations build reserves to decommission or replace software. Equally few design in for this reality. We will begin to see this as all the systems created for the so-called "Year 2000 Problem" require replacement. U.S. productivity growth could fall off a cliff.

"There's no other major item most of us own that is as confusing, unpredictable and unreliable as our personal computers."[15] It is not of course the computer—the hardware usually works well—it is the software. In general, anything really is possible with software. That is the problem. I have seen managers ask countless times if something can be done. The truthful answer is almost always

---

[12]Monica Lam, Professor of Computer Science, Stanford University, private communication, November 7, 2003.

[13]When Windows 2000 came out, it was estimated to contain upward of 30 million lines of code. Brent Schlender, "The Edison of the Internet," *Fortune*, February 15, 1999.

[14]Brent Schlender, "The Edison of the Internet," *Fortune*, February 15, 1999.

[15]Walter Mossberg, "Mossberg's Mailbox," *Wall Street Journal*, June 24, 2004.

yes. The problem is with the question: Yes it can be done, but will the software work reliably at scale over time at what expense and at what impact on further changes and replacement? It is easy to be a hero by making quick and dirty changes that leave your successor with enormous problems that are untraceable back to the "hero." Software is seldom managed well in my experience because the people managing it seldom have experience doing so.

SOFTWARE STACK

Software is a *stack* composed of multiple layers of software created by different teams over time for different purposes and brought together for the unique task at hand. Individual elements of software do not matter—what matters is the entire stack. The stack of software runs the computer and begins with a small piece of code called the *kernel.* The kernel allocates resources on the computer among the several claims for those resources. Simple computers and all the early computers did not have a kernel because they ran only one program at a time. The next layer is the *operating system*, which includes the kernel and other basic software on which all other software operates and to which all other software is written. For desktop computers, the windows manager presents the display and implements the user interface. For Windows and Mac it is tightly bound to the operating system, but for UNIX variants and Linux it is not. The window manager today can be larger and more complicated than the operating system. The next layer, *middleware*, hides the window manager and operating system from the application and provides a higher level of abstraction in which to program. Finally, there are the *applications*, which perform a useful purpose and consist of business logic, a user interface, and interactions with other systems.

When something goes right or something goes wrong with a computer, it is the entire software stack that is operating. If one piece of the hundred million lines of code is not working correctly, it affects the entire stack, so one of the biggest challenges is to view the stack as a whole, rather than the individual components. Except for some embedded systems, all running software stacks are unique. No two are alike except by pure, random chance.

Researchers have known for years that there are powerful heuristics that play on the software development process. The very best software developers are orders of magnitude better than the average, and the software stack leverages these wizards so that mortals can write effective software. There are only a very limited number of software developers worldwide at this high level that end up writing the basic kernel of the software we use.

The late Maurice Halstead in his book, *Elements of Software Science*,[16] explored at length how one attribute of the human brain, modestly called Halstead

[16]Maurice H. Halstead, *Elements of Software Science*, New York: Elsevier North Holland, 1977.

length, drove programming ability.[17] No one can solve a problem he or she cannot understand, and Halstead length defines how complicated a problem a person can understand. He found an average of about 250, which in his metrics was about one page of FORTRAN, but wizard programmers appear to have Halstead lengths over 65,000.

Any software stack is defined not only by its specifications, but also by its embedded errors and by its undocumented features. Many software developers are forced to build new software that is "bug-for-bug" compatible, as preexisting software bugs must be recognized and replicated for the new software to work. Even the best software contains errors and "even experienced engineers on average will inject a defect every nine to ten lines of code."

KNOWLEDGE STACK

Applications knowledge converts business rules and practice into algorithms and processes that can be implemented in software. People who do this must understand both the application as well as how to create software. Gaining and sustaining this application knowledge, therefore, can be a monumental challenge. Many large organizations today, for example, face a challenge in re-implementing their original software because many of the software developers who originally wrote the software have retired. Since rules for doing business are encoded into software, a firm may find that it no longer understands the details of how it operates.

Systems knowledge is the ability to create a working system to operate the software reliably and at scale. "Computer" knowledge is the ability to make the movement of ones and zeros a system. Few people ever have all this knowledge, so software is a uniquely team activity. At the same time, this also means that just adding additional software developers will not necessarily create the complementarities needed to combine both systems knowledge with computer knowledge successfully. Making the task of writing applications software uniquely complicated is the reality that software is a stack composed of multiple layers of software, usually created by different teams over times for different purposes and brought together for the unique task at hand.

SOFTWARE DEVELOPMENT

The way the industry has dealt with the complexity of software problems in the past 30 years is by introducing increasing levels of abstraction to the software

[17]An annotated bibliography of the psychology of programming compiled by Tim Matteson can be accessed at <*http://www.cise.ufl.edu/research/ParallelPatterns/PatternLanguage/Background/Psychology/Psych-bibliography.htm*>.

stack as you move up it. The increasing abstraction allows for easier portability across systems and expands the pool of people able to create software. The higher the level of abstraction generally the less efficient is the software; much of the gain in computing power historically has been used to create additional abstraction levels.

The labor to create each of the layers in the software stack is highly specialized (except that application builders usually also handle the user interface and the application error handling) and becomes increasingly rare as you go down the software stack. The reason that developers introduce abstraction is to allow for an increase in the number of workers that can work on a particular software project. While millions can write applications in Microsoft Visual Basic, the global labor pool that can write at the operating system kernel level is measured in the hundreds.

SYSTEMS INTEGRATION

The major costs in making a system operational are in configuration, testing, and tuning, a process inherently messy and difficult to manage. Packaged software almost never represents more than 5 percent of the total project cost.[18] While this 5 percent is tracked in the budget, 95 percent is expensed to G&A or divided into fractions and into different expenses that make the project extremely difficult to measure. This process also takes many more people (1 designer, 10 coders, and 100 testers would be a good starting point), but managers tend to allow too few resources and time for this step. The vast majority of needed resources for any corporate information system are non-system. Furthermore, when software is tested, the complete software stack is being tested and errors anywhere may show up for the first time.

Following the initial configuration, test and tune, any system has to be maintained over time. The lower portions of the software stack change almost daily, and so-called "patch management" becomes a significant task. Most readers see this process in something called Microsoft Windows Update. There are costs just in managing the computers and networks that operate the system. The trade-offs between operation and maintenance and development and replacement are huge, with many opportunities to save near-term expense for long-term costs. Thus, from the engineering perspective, the immense difficulty of measuring or predicting "flexibility" or "maintainability" or "reusability" during development means that development managers are often unable to steer a prudent course between over-investment, which creates unused flexibility and slows down development,

[18]Packaged software refers to an application program or collection of programs developed to meet the needs of a variety of users, rather than custom designed for a specific user or company. Packaged software is sold to the general public in shrink-wrapped boxes or, increasingly, by downloads over the Internet.

and under-investment, which can lead to unhappy "time-bomb" failures when small changes are made during maintenance.

SOFTWARE IS OFTEN MISCAST

Because of the lack of understanding about software discussed above, software is often miscast. Most of the existing models account for software in commercial and national accounts in ways that rely on flawed analogies to machines. Forty years ago, software was given away for free, but it has now evolved from a minor part of the value to nearly all the value in the system and value is often totally unrelated to cost.

To begin to understand the economics of software, we must understand that software affects production not as a factor of production like labor or capital, although software professionals and computing resources may be, but by defining the production function. While in general software improves the efficiency of the production function (i.e., you can produce more from the same inputs), it does so by increasing complexity, creating a probabilistic risk of catastrophic loss and raising the cost and risk of further change.

Software takes on economic value in the context of a system. In the case of desktop computers, lower levels of the software stack are bought and sold in the market and can be valued by their market price, although the natural monopoly aspects of these layers induces distortion. Because software is a large business, these metrics are interesting, but they do not capture at all the overall productivity implications for the economy.

Economists are not alone at all in miscasting software. The accounting and legal professions have both cast software in ways that fit their preexisting mental models. There is nothing malicious here at all by anyone. Software remains a mysterious black box to nearly all, and those who do understand its black arts are often enigmatic, unusual, even difficult people. Software is a unique human activity in which knowledge is congealed and given external form processable by both humans and machines, and it is completely dependent on a small labor pool of highly-talented designers and subject to little-understood but very powerful constraints. ***This knowledge must be, but never is, represented exactly, and errors cause damage but are today inevitable.***

Fine software is like fine wine: there is no known way to produce error-free software except time. IBM mainframes now often run years without failure, but that software is more than 30 years old. The man who led the effort to create the operating system for IBM mainframes was Fred Brooks, and his book on software engineering, *The Mythical Man Month*,[19] remains seminal today. Among

[19]Frederick Brooks, *The Mythical Man Month: Essays on Software Engineering*, Reading, MA: Addison-Wesley Publishing Company, 1975, pg. 116.

his key tenets was the Plan to Throw One Away rule: the only way to design a system is to build it. In my personal experience, the wisdom of this adage is manifestly obvious, but not to most executives who see it as wasteful.

The only complete specification for any software stack is the stack itself, so the only way to design a system is to build it. Software designers understand that a specification without a reference implementation is meaningless. This is why compatibility and portability have been so elusive. The published specifications for Java 2 Enterprise Edition, the standard today for building corporate information systems, measure slightly over a meter high.

ECONOMICS OF SOFTWARE

Because of a lack of a good economic model of software, measurement remains enigmatic. At the root, the declining costs of silicon have enabled massive change, but it is software that converts this change into productivity and quality improvements. Unfortunately, we manage and account for software largely as a period expense, when in fact it is a balance sheet transaction adding both assets and liabilities. The real data needed to understand what is happening simply do not exist in firm records.

Some of the complications of measuring the value of software start with the fact that most people think of software as the programming phase while, in reality, programming costs are usually less than 10 percent of the total system cost. The systems design is an intangible asset whose value grows and shrinks with the operations and prospects for the business. The implementation is an intangible asset of potentially enormous value, as finding even one way to make a design work reliably at scale is not assured, but it also is a liability, for it must be maintained and evolved as the business changes over time.[20]

The costs of maintaining and modifying commercial software increase over time and increase with the amount of accumulated change.[21] Change is required by the rapid change in the foundation platform as well as by the evolution of the environment within which the software operates, and the requirements and needs it fulfills. As mentioned above, software on average has a useful life of 7 to 10 years, after which time it must be replaced. This comes from two forces: accumulated change and platform evolution.

Every software veteran understands that a manager can choose only two of schedule, cost, and features, where features includes quality and performance. In

[20]Platforms and infrastructure change very rapidly also.

[21]An alternate view is that much of maintenance activity, which is treated as intermediate consumption—and not investment—in the national accounts, may have effects on productivity. Nevertheless, it is not counted as investment. Similarly, although digital telephone switches undergo continuing software changes, the expenditures for these changes are not treated as investment, but are part of intermediate consumption.

turn, quality, performance and capabilities all trade off. The most important way to cut costs and time and increase features is to improve the quality of the developers (which can be done to some degree by increasing the quality and power of their tools). Adding more people is the surest way to make a late software project later. An old adage is that software is 90 percent complete for 90 percent of the development cycle.

Putnam studied these tradeoffs at GE decades ago and found that halving the schedule required a 16-fold increase in total staffing, everything else being the same.[22] He also found that increasing complexity had massive impact on total cost: 10 percent increase in complexity required a threefold increase in the number of developers working simultaneously. Software design is at least 90 percent of whether software is good or bad, and the quality of the designer is at least 90 percent of the quality of the design.

PUBLIC POLICY ISSUES

Although, to my knowledge at least, no major politician has yet been asked for a software agenda, public policy issues abound:

- **Are we as a nation investing adequately in the systems that improve productivity for the economy?** Indeed, how much are we investing in such systems and how is that amount trending? Does the United States face a productivity cliff in the next few years when the systems installed for the year 2000 age and prove difficult to replace or upgrade? What public policies encourage the creation of such productivity-enhancing systems? No data currently exist to answer these questions.

- **Do public corporations properly report their investments in software and their resulting expenses?** Can investors accurately understand corporate performance and outlooks without better knowledge of the software systems that operate modern corporations and products? Telecommunications companies' main challenge today is implementing billing systems whose software can effectively communicate with other systems.

- **Do traditional public policies on competition work when applied to software-related industries?** Is software really so different that traditional economic regulations and remedies no longer work? Software changes so fast that by the time traditional public policies work, the issues have already changed.

- **Do we educate properly in our schools and universities given the current and growing importance of software?** What should an educated person

[22]This may not be typical today.

know about software? Corporate reporting and control systems are software systems. What should the professionals charged with operating and inspecting those systems know about software?

- **What should be our policy on software and business methods patents?** Recently, the European Union adopted a fundamentally different course from the United States.[23] One recurring issue in the United States is that new patents are coming out on ideas from 30 years ago, but the prior art is not readily available.

- **What is the proper level of security for public and private systems?** How do we achieve that? What are the externalities of security? Are special incentives required? There is a proposal in some U.S. government agencies that would require companies to certify in their public reporting the security level of all their key systems, particularly those that are part of the critical infrastructure. These companies would then have to take responsibility for the security of these systems. The worms that have been circulating lately have the potential to infiltrate home computers and create an unstoppable attack on vital infrastructure as 10-20 million broadband connected homes would be simultaneously attacking infrastructure.

- **What is happening to software jobs?** Should our policies change? What are the implications of the apparent migration to India and China? Is the United States simply getting lower-cost labor that will enable further advancements in the United States or are these jobs the tip of the value chain that will eventually make sustaining a software system difficult?

- **What export controls make sense for software?** Most software is dual use in the sense that it can be used for military and non-military purposes.

- **Should we encourage or enable open-source software?** Is this the right economic model given the nature of talented software people? What is the proper legal environment for open-source software? Should governments mandate its use on the fundamental principle that the data of the people cannot be captive to any proprietary vendor ever, a principle increasingly voiced and occasionally enacted, most recently by Munich?[24] Is open source really the best economic model for software? Is open-source software more or less reliable and secure? Source code, instead of being the most valuable part of software, is the least valuable part of software and giving it away may be a good strategy to promote new innovation.

---

[23]Paul Meller, "European Parliament Votes to Limit Software Patents," *New York Times*, September 25, 2003.

[24]Stephen Pritchard, "Munich Makes the Move," *Financial Times*, October 15, 2003.

• **Should strict liability apply at all to software?** Should software professionals be licensed? Software traditionally has been sold in the United States on an as-is basis, although that is now being tested in the courts. While process, programming language, frameworks, architecture, tools and infrastructure have all evolved considerably in the last 5 years, there is still no fundamentally new way of writing software. Bill Joy has highlighted this issue by observing that almost all software today is created using variants of what he would call antique technologies. [25] If the U.S. could find a new to way to write software, it would have an enormous positive impact on the value equation for the U.S. economy.

• **How do we get the right investment into fundamental software technology?** We have little visibility into how much we are investing.

Although this list is not exhaustive by any means, it is apparent that the public policy agenda is rich. Software pervades our economy and our society and should, not surprisingly, also pervade our policy agenda. As we move forward, the awareness of software's importance must be raised and a dialogue must be started in the hopes of better understanding the economics of software.

APPENDIX: THE SOFTWARE PROCESS

Designing the software and specifying its architecture is the first and most important task.

• *Rule one is to hire great people and make one of them the chief architect.* Everyone else has to play a supporting role. Paranoia is a desirable attribute for the chief architect. Experience proves that the only people capable of hiring well are great software developers themselves.

• *Rule two is to encapsulate complexity as much as possible.* The costs of spreading complexity are enormous. This explains why changing software in ways not anticipated by the design is so costly because doing so inherently spreads complexity. Complexity is viral.

• *Rule three is to write at the highest abstraction layer possible.*

• *Rule four is to use the best tools you can find.* This is harder than it seems. Study after study showed that writing Java code could be up to ten times more efficient than writing C code,[26] but many developers did not like the constraints

[25]Interview with Brent Schlender in *Fortune*, September 15, 2003.

[26]James Gosling, Bill Joy, and Guy Steele, *The Java (TM) Language Specification*, Reading, MA: Addison-Wesley, 1996.

that Java imposed. In addition, the cost-plus economics of software development sometimes take away the productivity gains provided by good tools.

Properly encapsulating complexity can greatly reduce the cost and increase the life of software. This is why object-oriented programming and modern programming languages such as Java are so important.[27] An error should affect only the item with the error: memory leaks being the counterexample. When your personal computer suddenly and inexplicably ceases to work, the vast majority of the time you are probably the victim of a memory leak: some application somewhere, maybe one you did not even know was running, failed to manage its use of system memory properly leading to system failure.[28]

The phase most people think of as software is the implementation or purchase of programming. A bad implementation of a great design is probably mediocre software, but a great implementation of a bad design is probably miserable software. The costs here are usually less than 10 percent of the total system cost. "Technical feasibility assured," the accounting standard for capitalization, occurs in this phase.[29]

This phase often appears to be mechanical, like constructing a house. I was once told that my economics colleagues considered software to be like the glass blower was to the chemist: that is, of no intellectual value. Industry slang often refers to something as SMOP, a simple matter of programming, but programming in general is never simple. The phrase in fact is meant as a pun.

It is usually unclear when design ends and implementation begins. The worse the software the more unclear you can be sure it was when it was created. More importantly, it is also unclear when implementation ends. Software always has errors. Developers do not measure absolute errors. Instead, they measure the rate at which new errors are discovered.

REFERENCES

Brooks, Frederick. 1975. *The Mythical Man Month: Essays on Software Engineering*. Reading, MA: Addison-Wesley Publishing Company.

Computerworld. 1999. "Generally Accepted Accounting Principles." November 29.

Executive Office of the President. 2003. *The National Strategy to Secure Cyberspace: Cyberspace Threats and Vulnerabilities*. February.

Hagel, J. and A.G. Armstrong. 1997. *Net Gain*. Cambridge, MA: Harvard Business School Press.

Halstead, Maurice H. 1977. *Elements of Software Science*. New York: Elsevier North Holland.

---

[27]A principal innovation of Java, C#, and Ada95 over C and C++ is the language-enforced strong encapsulation of data.

[28]This was what caused the recent problems with the two Mars rovers. In practice, most computer freezes are caused by deadlocks or race-induced data corruption.

[29]"Generally Accepted Accounting Principles," *Computerworld*, November 29, 1999.

Jorgenson, Dale W. 2002. "The Promise of Growth in the Information Age." The Conference Board, Annual Essay.

Kessler, Michelle. 2002. "Computer Majors Down Amid Tech Bust." *USA Today* October 8.

Lam, Monica. 2003. Private communication. November 7.

Maney, Kevin. 2003. "Music Industry Doesn't Know What Else To Do As It Lashes Out at File-sharing." *USA Today* September 9.

Meller, Paul. 2003. "European Parliament Votes to Limit Software Patents." *New York Times* September 25.

Mossberg, Walter. 2004. "Mossberg's Mailbox." *Wall Street Journal* June 24.

National Research Council. 2005. *Deconstructing the Computer.* Dale W. Jorgenson and Charles W. Wessner, eds. Washington, D.C.: The National Academies Press.

Oliner, Stephen and Daniel Sichel. 2000. "The Resurgence of Growth in the Late 1990s: Is Information Technology the Story?" *Journal of Economic Perspectives* 14(4) Fall.

Organisation for Economic Co-operation and Development. 2002. *Information Technology Outlook 2002—The Software Sector.* Paris: Organisation for Economic Co-operation and Development.

Organisation for Economic Co-operation and Development. 2002. *Strengthening the Knowledge-based Economy.* Paris: Organisation for Economic Co-operation and Development.

Organisation for Economic Co-operation and Development. 2003. *ICT and Economic Growth: Evidence from OECD Countries, Industries and Firms.* Paris: Organisation for Economic Co-operation and Development.

Poulsen, Kevin. 2004. "Software Bug Contributed to Blackout." *Security Focus* February 11.

Pritchard, Stephen. 2003. "Munich Makes the Move." *Financial Times* October 15.

Schlender, Brent. 1999. "The Edison of the Internet." *Fortune* February 15.

Shuster, Loren. 2002. "Global Gaming Industry Now a Whopping $35 Billion Market." *Compiler* July 2002.

IV

APPENDIXES

Appendix A

Biographies of Speakers*

GREG BEAMS

Greg Beams is a partner in Ernst & Young's Washington, D.C. practice working within the firm's Attest practice. Greg is in his 16th year with Ernst & Young and focuses primarily on clients within the software and Internet industries. Greg has extensive experience in dealing with software company accounting and reporting issues, both for domestic and international software companies in the public and private sector. Greg graduated from Central Washington University with B.S.s in accounting and finance and is a licensed CPA in the State of Virginia.

ERNST R. BERNDT

Ernst R. Berndt is the Louis B. Seley Professor of Applied Economics at the MIT Sloan School of Management. Professor Berndt also serves as Director of the National Bureau of Economic Research Program on Technological Progress and Productivity Measurement, and is Adjunct Professor of Health Care Policy and Management at the Harvard Medical School. Much of Prof. Berndt's research

*As of February 2004.

focuses on the measurement and identification of factors affecting the pricing and diffusion of new technologies, such as PC hardware, PC software, and novel medical innovations.

Dr. Berndt received his B.A. (Honors) degree in economics from Valparaiso University in 1968, where he was a Christ College Scholar, an M.S. (1971) and Ph.D. (1972) in economics from the University of Wisconsin-Madison, and an honorary doctorate (1991) from Uppsala University in Sweden. Currently Dr. Berndt serves as Chair of the Federal Economic Statistics Advisory Committee, is a panel review member of the Methodology, Measurement and Statistics program at the National Science Foundation, and is on Independent Detail from the U.S. Food and Drug Administration.

WILLIAM B. BONVILLIAN

William Bonvillian is the Legislative Director and Chief Counsel to Senator Joseph I. Lieberman (D-CT). Prior to his work on Capitol Hill, he was a partner at both the law firms of Jenner & Block as well as Brown & Roady. Early in his career, he served as the Deputy Assistant Secretary and Director of Congressional Affairs at the U.S. Department of Transportation. His recent articles include, "Organizing Science and Technology for Homeland Security," in *Issues in Science and Technology,* and "Science at a Crossroads," published in *Technology in Society*. His current legislative efforts at Senator Lieberman's office include science and technology policy and innovation issues.

Mr. Bonvillian is married to Janis Ann Sposato and has two children. He received his B.A. from Columbia University; his M.A.R. from Yale University; and his J.D. from Columbia Law School where he also served on the Board of Editors for the Columbia Law Review. He is a member of the Connecticut Bar, the District of Columbia Bar, and the U.S. Supreme Court Bar.

KENNETH FLAMM

Kenneth Flamm is Professor and Dean Rusk Chair in International Affairs at the LBJ School of Public Affairs at the University of Texas-Austin. He is a 1973 honors graduate of Stanford University and received a Ph.D. in economics from MIT in 1979. From 1993 to 1995, Dr. Flamm served as Principal Deputy Assistant Secretary of Defense for Economic Security and Special Assistant to the Deputy Secretary of Defense for Dual Use Technology Policy. Prior to and after his service at the Defense Department, he spent 11 years as a Senior Fellow in the Foreign Policy Studies Program at Brookings. Dr. Flamm has been a professor of economics at the Instituto Tecnológico A. de México in Mexico City, the University of Massachusetts, and George Washington University.

Dr. Flamm currently directs the LBJ School's Technology and Public Policy Program, and directs externally funded research projects on "Internet Use in

Developing and Industrializing Countries," "The Economics of Fair Use," and "Determinants of Internet Use in U.S. Households," and has recently initiated a new project on "Exploring the Digital Divide: Regional Differences in Patterns of Internet Use in the U.S." He continues to work with the semiconductor industry research consortium International SEMATECH, and is building a return-on-investment-based prototype to add economic logic to SEMATECH's industry investment model. He also is a member of the National Academy of Science's Panel on the Future of Supercomputing, and its Steering Group on Measuring and Sustaining the New Economy. He has served as member and chair of the NATO Science Committee's Panel for Science and Technology Policy and Organization, and as a member of the Federal Networking Council Advisory Committee, the Organisation for Economic Co-operation and Development's Expert Working Party on High Performance Computers and Communications, and various advisory committees and study groups of the National Science Foundation, the Council on Foreign Relations, the Defense Science Board, and the U.S. Congress' Office of Technology Assessment, and as a consultant to government agencies, international organizations, and private corporations.

Dr. Flamm is the author of numerous articles and books on the economic impacts of technological innovation in a variety of high-technology industries. Among the latter are *Mismanaged Trade? Strategic Policy and the Semiconductor Industry* (1996), *Changing the Rules: Technological Change, International Competition, and Regulation in Communications* (ed., with Robert Crandell, 1989), *Creating the Computer* (1988), and *Targeting the Computer* (1987). Recent work by Flamm has focused on measurement of the economic impact of the semiconductor industry on the U.S. economy, analyzing the economic determinants of Internet use by households, and assessing the economic impacts of Internet use in key applications.

JACK HARDING

Jack Harding brings 20 years of executive management experience in the electronics industry to eSilicon and serves as chairman, president, and chief executive officer (CEO). Prior to eSilicon, Mr. Harding served as President and CEO of Cadence Design Systems. During his tenure Cadence was the world's largest supplier of electronic design software. From 1994 to 1997, Mr. Harding was President and CEO of Cooper & Chyan Technology, which was acquired by Cadence in 1997. Harding served as executive vice president of Zycad Corporation from 1984 to 1994. He began his career at IBM. Mr. Harding holds a B.A. in economics and chemistry from Drew University, where he is a vice chairman of the Board of Trustees. Harding is a Senior Fellow at the Institute for Development Strategies, and a member of the Board of Visitors for the School of Public and Environmental Affairs, Indiana University. He is a member of the Council on Competitiveness, a Washington, D.C.-based organization of Fortune 500 CEOs and university presidents dedicated to the global competitiveness of the United States. Mr. Harding also serves on the Board of Directors of Marimba, Incorporated. He is a frequent lecturer on innovation and entrepreneurship.

RONIL HIRA

Ronil Hira is an assistant professor of Public Policy at Rochester Institute of Technology. He specializes in engineering workforce issues and innovation policy. He completed his post-doctoral fellowship at Columbia University's Center for Science, Policy, and Outcomes. Dr. Hira holds a Ph.D. in public policy from George Mason University (GMU), an M.S. in electrical engineering also from GMU, and a B.S. in electrical engineering from Carnegie Mellon University. He has been a consultant for the Rand Corporation, Deloitte & Touche, and Newport News Shipbuilding. He testified before Congress on the implications of offshore outsourcing on innovation and engineering careers. Dr. Hira was previously a program manager at the National Institute of Standards and Technology. He is a licensed professional engineer and is currently Chair of the Career and Workforce Policy Committee of IEEE-USA.

DALE W. JORGENSON

Dale Jorgenson is the Samuel W. Morris Professor of Economics at Harvard University. He has been a Professor in the Department of Economics at Harvard since 1969 and Director of the Program on Technology and Economic Policy at the Kennedy School of Government since 1984. He served as Chairman of the Department of Economics from 1994 to 1997. Dr. Jorgenson received his Ph.D. in economics from Harvard in 1959 and his B.A. in economics from Reed College in Portland, Oregon, in 1955.

Dr. Jorgenson was elected to membership in the American Philosophical Society in 1998, the Royal Swedish Academy of Sciences in 1989, the U.S. National Academy of Sciences in 1978, and the American Academy of Arts and Sciences in 1969. He was elected to Fellowship in the American Association for the Advancement of Science in 1982, the American Statistical Association in 1965, and the Econometric Society in 1964. Uppsala University and the University of Oslo awarded him honorary doctorates in 1991.

Dr. Jorgenson is president of the American Economic Association. He has been a member of the Board on Science, Technology, and Economic Policy of the National Research Council since 1991 and was appointed to be Chairman of the Board in 1998. He is also Chairman of Section 54, Economic Sciences, of the National Academy of Sciences. He served as President of the Econometric Society in 1987.

Dr. Jorgenson is the author of more than 200 articles and the author and editor of 20 books in economics. The MIT Press, beginning in 1995, has published his collected papers in 9 volumes. The most recent volume, *Econometrics and Producer Behavior*, was published in 2000.

Prior to Dr. Jorgenson's appointment at Harvard he was Professor of Economics at the University of California, Berkeley, where he taught from 1959 to 1969. He has been Visiting Professor of Economics at Stanford University and the Hebrew University of Jerusalem and Visiting Professor of Statistics at Oxford University. He has also served as Ford Foundation Research Professor of Economics at the University of Chicago.

Forty-two economists have collaborated with Dr. Jorgenson on published research. An important feature of Dr. Jorgenson's research program has been collaboration with students in economics at Berkeley and Harvard, mainly through the supervision of doctoral research. This collaboration has often been the outgrowth of a student's dissertation research and has led to subsequent joint publications. Many of his former students are professors at leading academic institutions in the United States and abroad and several occupy endowed chairs.

MONICA LAM

Monica Lam received a B.Sc. from the University of British Columbia in 1980 and a Ph.D. in computer science from Carnegie Mellon University in 1987. She joined the faculty of Computer Science at Stanford in 1988, where she is now a professor. Her research interests are in systems: program analyses, operating systems, and architectures.

Dr. Lam's current research projects focus on making computing and programming easier. The Collective project she leads is developing a computing utility, based on the concept of virtual appliances, whose goals are to simplify system administration and support user mobility. Her program analysis group has recently developed a number of practical programming tools including a static memory leak detector called Clouseau, a dynamic bounds-checker called CRED, and a dynamic error detection and diagnosis tool called DIDUCE.

Dr. Lam led the SUIF (Stanford University Intermediate Format) Compiler project, which produced a widely used compiler infrastructure known for its locality optimizations and interprocedural parallelization. Many of the compiler techniques she developed have been adopted by the industry. Her other research projects included the architecture and compiler for the CMU Warp machine, a systolic array of VLIW processors, and the Stanford DASH distributed shared memory machine. In 1998, she took a sabbatical leave from Stanford to help start Tensilica Inc., a company that specializes in configurable processor cores.

Honors for Dr. Lam's research work at Stanford include an NSF Young Investigator award, an ACM Most Influential Programming Language Design and Implementation Paper Award, and an ACM SIGSOFT Distinguished Paper Award. She chaired the ACM SIGPLAN Programming Languages Design and Implementation Conference in 2000, served on the Editorial Board of ACM

Transactions on Computer Systems, and numerous conference program committees including ASPLOS, ISCA, PLDI, POPL, and SOSP.

SHELLY C. LUISI

Shelly Luisi is a senior associate chief accountant in the Office of the Chief Accountant (OCA) of the U.S. Securities and Exchange Commission and co-leader of OCA's technical accounting staff. As the SEC's technical body on U.S. and international financial reporting matters, OCA is responsible for overseeing the activities of the Financial Accounting Standards Board and its designees, monitoring the activities of international accounting standard setters, and consulting with registrants, auditors, and SEC staff regarding complex financial reporting issues. Ms. Luisi is responsible for the processes related to interpretations of accounting standards, including oversight of the activities of the Emerging Issues Task Force, analysis of the activities of the International Financial Reporting Interpretations Committee, and consultations on financial reporting issues.

Prior to joining the SEC in July 2000, Ms. Luisi worked in the Atlanta and Salt Lake City offices of Ernst & Young and the Las Vegas office of KPMG, as well as in industry as Vice President, Accounting and Information Systems of an SEC registrant.

Ms. Luisi earned a B.S. degree with a major in accounting in 1990 and an M.Ac. degree with a focus on information systems in 1991, both from Florida State University. She is a CPA, licensed by the state of Georgia.

MARK B. MYERS

Mark B. Myers is visiting executive professor in the Management Department at the Wharton Business School, the University of Pennsylvania. His research interests include identifying emerging markets and technologies to enable growth in new and existing companies with special emphases on technology identification and selection, product development, and technology competencies. Mark Myers serves on the Science, Technology, and Economic Policy Board of the National Research Council and currently co-chairs with Richard Levin, the President of Yale, the National Research Council's study of "Intellectual Property in the Knowledge Based Economy."

Dr. Myers retired from the Xerox Corporation at the beginning of 2000, after a 36-year career in its research and development organizations. Myers was the senior vice president in charge of corporate research, advanced development, systems architecture, and corporate engineering from 1992 to 2000. His responsibilities included the corporate research centers: PARC in Palo Alto, California; Webster Center for Research & Technology near Rochester, New York; Xerox Research Centre of Canada, Mississauga, Ontario; and the Xerox Research Centre of Europe in Cambridge, UK, and Grenoble, France. During this period he was a member of the senior management committee in charge of the strategic direction setting of the company.

Dr. Myers is chairman of the board of trustees of Earlham College and has held visiting faculty positions at the University of Rochester and at Stanford University. He holds a B.S. from Earlham College and a Ph.D. from Pennsylvania State University.

DIRK PILAT

Dirk Pilat is a senior economist with responsibilities for work on Productivity, Growth, and Firm-level Analysis at the Organisation for Economic Co-operation and Development (OECD) Directorate for Science, Technology and Industry. He holds a Ph.D. in economics from the University of Groningen in the Netherlands, where he was also a research fellow. He joined the OECD in 1994, working on unemployment, regulatory reform, product market competition, and economic growth. In recent years, his work has focused on economic growth and issues related to the "new economy," during which he contributed to OECD reports such as *ICT and Economic Growth: Evidence from OECD Countries, Industries and Firms* (2003), *Seizing the Benefits of ICT in a Digital Economy* (2003), *The New Economy—Beyond the Hype* (2001), and *A New Economy? The Changing Role of Innovation and Information Technology in Growth* (2000). Since 1998, he has been a member of the editorial board of OECD Economic Studies and of the Review of Income and Wealth.

WILLIAM J. RADUCHEL

William J. Raduchel provides strategic advice and consulting to a number of technology-related companies including America Online (AOL), Chordiant Software, where he is a director, Myriad International, Silicon Image, where he also serves as chairman of PanelLink Cinema Partners PLC, Hyperspace Communications, and Wild Tangent.

Through 2002 Dr. Raduchel was executive vice president and chief technology officer of AOL Time Warner, Inc, after being senior vice president and chief technology officer of AOL. Infoworld named him CTO of the year in 2001. Dr. Raduchel joined AOL in September 1999 from Sun Microsystems, Inc, where he was chief strategy officer and a member of its executive committee. In his 11 years at Sun, he also served as chief information officer, chief financial officer, acting vice president of human resources and vice president of corporate planning and development and oversaw relationships with the major Japanese partners. He was recognized separately as CIO of the year and as best CFO in the computer industry.

In addition, Dr. Raduchel has held senior executive roles at Xerox Corporation and McGraw-Hill, Inc. He is a member of the National Advisory Board for the Salvation Army, the Board of Directors of In2Books, the National Research

Council Committee on Internet Navigation and Domain Name Services and the Board on Science, Technology, and Economic Policy of the National Research Council. He has several issued and pending patents.

After attending Michigan Technological University, which gave him an honorary doctorate in 2002, Dr. Raduchel received his undergraduate degree in economics from Michigan State University, and earned his A.M. and Ph.D. degrees in economics at Harvard. In both the fall and spring of 2003 he was the Castle Lecturer on Computer Science at the U.S. Military Academy at West Point.

WAYNE ROSING

Wayne Rosing brought to Google more than 30 years of engineering and research experience at some of Silicon Valley's most respected companies. That experience in forming high-performance engineering teams has served him well at Google where he is responsible for a staff of more than 100 technical professionals working in small teams on multiple projects. Mr. Rosing joined Google from Caere Corporation, where his most recent position was chief technology officer and vice president of Engineering. Mr. Rosing managed all engineering for Caere's optical character recognition (OCR) product lines and was the driving force behind the acquisition of the comprehensive forms application Omniform, which became one of Caere's key products.

Prior to joining Caere, Mr. Rosing served as president of FirstPerson, Inc., a wholly owned subsidiary of Sun Microsystems. While at FirstPerson, Mr. Rosing headed the team that developed the technology base for Java. That success was preceded by his founding of Sun Microsystems Laboratories, which grew to more than 100 researchers under his leadership. Mr. Rosing worked at Sun Microsystems in various executive positions from 1985 through 1994. Earlier in his career, Mr. Rosing was director of engineering for the Apple Computer Lisa and Apple II divisions and held management positions at Digital Equipment Corporation and Data General.

ANTHONY SCOTT

As chief information technology officer at General Motors, Mr. Scott is responsible for defining the information technology computing and telecommunications architecture and standards across all of the company's business globally.

Mr. Scott joined General Motors from Bristol-Myers Squibb where he was vice president, Information Management for the Shared Services Group. He has also held positions as senior director, Technology Knowledge Organization with Price Waterhouse; vice president of Engineering with Uniteq Application Systems; and manager, Worldwide Information Resources with Sun Microsystems.

Mr. Scott completed a B.S. degree from the University of San Francisco in information systems management and a Juris Doctorate from Santa Clara University. He, his wife, and two teenage sons are from New Jersey.

JAMES SOCAS

James Socas currently serves on the staff of the U.S. Senate Banking Committee with responsibility for the Economic Policy Subcommittee and works closely with U.S. Senator Chuck Schumer. He has focused closely on issues relating to the "offshoring" of U.S. jobs and trade relations with China. Prior to joining the Senate staff, he was a managing director at Credit Suisse First Boston (CSFB), responsible for software industry clients and transactions, and served in a similar capacity at Donaldson Lufkin & Jenrette, before they were acquired by CSFB in 2000. Mr. Socas also served as Assistant to the President for Reston-Virginia based Perot Systems, an IT-services company, and helped start a small clinical software business. He is a passionate believer in the importance of high-tech industries to the future growth and job creation in the country. Mr. Socas is a graduate of the Harvard Business School and the University of Virginia and lives in McLean, Virginia with his wife and two children.

HAL R. VARIAN

Hal R. Varian is the Class of 1944 Professor at the School of Information Management and Systems, the Haas School of Business, and the Department of Economics at the University of California (UC) at Berkeley.

He received his S.B. degree from MIT in 1969 and his M.A. (mathematics) and Ph.D. (economics) from UC Berkeley in 1973. He has taught at MIT, Stanford, Oxford, Michigan, and other universities around the world.

Dr. Varian is a fellow of the Guggenheim Foundation, the Econometric Society, and the American Academy of Arts and Sciences. He has served as co-editor of the *American Economic Review* and is on the editorial boards of several journals.

Professor Varian has published numerous papers in economic theory, industrial organization, financial economics, econometrics, and information economics. He is the author of two major economics textbooks which have been translated into 22 languages. His current research has been concerned with the economics of information technology and the information economy. He is the co-author of a bestselling book on business strategy, *Information Rules: A Strategic Guide to the Network Economy,* and writes a monthly column for the *The New York Times*.

KENNETH WALKER

Kenneth Walker currently serves as SonicWALL's director of Platform Evangelism where he is engaged in the development of a security ecosystem stretching beyond individual companies to help create layered security solutions for the broader market.

Prior to joining SonicWALL, Mr. Walker was most recently vice president of Technical Strategy for Philips Components, where he drove customer and technology innovation to deliver value across product lines. While at Philips, Walker served in a number of capacities in the display arena, managing teams on the cutting edge of LCD television and innovative display applications. Previously, Walker has held senior positions for Verano, Radius, and Apple Computer.

Mr. Walker has an M.S. in management and a B.S. in information and computer science from the Georgia Institute of Technology.

DAVID WASSHAUSEN

Dave Wasshausen is a supervisory economist with the Bureau of Economic Analysis (BEA). He is responsible for the estimates of private fixed investment in equipment and software, and of foreign transactions in the national income and product accounts.

Mr. Wasshausen co-authored and presented a paper entitled, "Information Processing Equipment and Software in the National Accounts," at an NBER/CRIW conference on "Measuring Capital in the New Economy." The paper is available on BEA's web site and will be published in a forthcoming CRIW volume.

Mr. Wasshausen received his M.A. in economics from The American University (1996) and his B.A. in economics from Miami University (1990). Mr. Wasshausen has been working as an economist in BEA's national accounts directorate since 1991.

ALAN G. WHITE

Alan G. White is a manager at Analysis Group, Inc. in Boston, a firm of economic, financial and strategy consultants. Dr. White specializes in the application of statistics, econometrics and applied microeconomics to litigation and general business problems. He has provided quantitative economic analyses in antitrust, intellectual property and complex business litigation. His work spans a wide variety of industries, including banking, computer hardware and software, vitamins, and agricultural products. His case work has included price measurement issues in a variety of industries, the application of quality-adjusted price methods, the evaluation of market conditions pertaining to class certification in the sales of genetically modified seeds, the examination of market conditions relevant to the effective function of a price-fixing cartel, and the evaluation of the cost of employees with various chronic pain conditions to employers. Dr. White has a B.A. (Honors) in mathematics and economics and a M. Litt. in economics from the University of Dublin, Trinity College, and a Ph.D. in economics from the University of British Columbia. Dr. White's research and publications have focused on economic measurement, including the construction of stock market indexes, and the theory and implementation of consumer and producer price indexes.

Appendix B

Participants List*

Ana Aizcorbe
U.S. Department of Commerce

Jeffrey Alexander
Washington CORE

John Alic

Tom Arrison
The National Academies

Hollis Bauer
Epsilon

Greg Beams
Ernst & Young

David Beede
U.S. Department of Commerce

Tabitha Benney
The National Academies

Andrew Bernat
Computing Research Association

Ernst R. Berndt
MIT Sloan School of Management

Richard Bissell
The National Academies

Mathieu Bonnet
Embassy of France

**Speakers in italics.*

William B. Bonvillian
Office of Senator Joseph Lieberman

Bryan Borlik
NACFAM

Betsy Brady
Microsoft

Katie Burns
International Economic Development Council

Matthew Bye
Federal Trade Commission

Erran Carmel
American University

Connie Chang
National Institute of Standards and Technology

Won Chang
U.S. Department of the Treasury

McAlister Clabaugh
The National Academies

Eileen Collins
Rutgers University

Michael Crowley
Akin Gump

Christopher Currens
National Institute of Standards and Technology

Donald Dalton
U.S. Department of Commerce

David Dierksheide
The National Academies

Carmen Donohue
Government Accountability Office

Michael Ehst
House Committee on Science

Kevin Finneran
The National Academies

Jonas Fjellman
Embassy of Sweden

Kenneth Flamm
University of Texas at Austin

Mark Fostek
Government Accountability Office

John Gardenier

Eric Garduño
International Intellectual Property Institute

Gerald Hane
Globalvation

Jack Harding
eSilicon Corporation

Meg Hardon
The Dutko Group

Christopher Hayter
Council on Competitiveness

Robert Hershey

Ronil Hira
Rochester Institute of Technology

Mike Holdway
U.S. Department of Labor

Eric Holloway
U.S. Department of Commerce

Thomas Howell
Dewey Ballantine, LLC

Kent Hughes
Woodrow Wilson Center

Virginia Hughes
Government Accountability Office

Olwen Huxley
House Committee on Science

Ken Jacobson
The National Academies

David Johnson
U.S. Department of Labor

Dale W. Jorgenson
Harvard University

Bradley Knox
Committee on Small Business
U.S. House of Representatives

Elka Koehler
Office of Senator Joseph Lieberman

Heikki Kotila
Embassy of Finland

Norman Kreisman
U.S. Department of Energy

Monica Lam
Stanford University

Rolf Lehming
National Science Foundation

Philip Leith
Queen's University of Belfast

Chan Lieu
Senate Committee on Commerce

Bill Long
Business Performance Research Associates

David Longstreet
Software Metrics

Shelly C. Luisi
Securities and Exchange Commission

Bill Marck
House Committee on Armed Services

Jeff Mayer
U.S. Department of Commerce

Richard McCormack
Manufacturing News

Hugh McElrath
Department of the Navy

Sean Mcilvain

Stephen A. Merrill
The National Academies

Egils Milbergs
Center for Accelerating Innovation

Sujata Millick
U.S. Department of Commerce

Sabrina Montes
U.S. Department of Commerce

Robert Morgan

Bill Morin
Applied Materials

Russell Moy
The National Academies

Carol Moylan
U.S. Department of Commerce

Kazue Muroi
Washington CORE

Jerry Murphy
Business-Higher Education Forum

Nathan Musick
Congressional Budget Office

Mark B. Myers
The Wharton School
University of Pennsylvania

John Nail
National Institute of Standards and Technology

Hironori Nakanishi
New Energy and Industrial Technology Development Organization (NEDO)

Nicolas Naudin
Embassy of France

Michael Nelson
International Business Machines

Vin O'Neill
IEEE-USA

Jayne Orthwein
National Institute of Standards and Technology

Dirk Pilat
Organisation for Economic Co-operation and Development

Erik Puskar
National Institute of Standards and Technology

William J. Raduchel

Lawrence Rausch
National Science Foundation

Larry Rosenblum
U.S. Department of Labor

Wayne Rosing
Google

John Sargent
U.S. Department of Commerce

Hsiu-Ming Saunders
The National Academies

Anthony Scott
General Motors

Stephanie Shipp
National Institute of Standards and Technology

Patricia Slocum
Government Accountability Office

James Socas
Senate Committee on Banking

Victor Souphom
Census Bureau

Gregory Tassey
National Institute of Standards and Technology

Suzy Tichenor
Council on Competitiveness

Hal R. Varian
University of California at Berkeley

Kenneth Walker
SonicWALL

David Wasshausen
Bureau of Economic Analysis

Philip Webre
Congressional Budget Office

Timothy Wedding
Government Accountability Office

Charles W. Wessner
The National Academies

Alan G. White
Analysis Group, Inc.

Feng Zhu
Harvard University

Appendix C

Selected Bibliography on Measuring and Sustaining the New Economy

Abel, Jaison R., Ernst R. Berndt, Alan G. White. 2003. "Price Indexes for Microsoft's Personal Computer Software Products." NBER Working Paper 9966.

Abel, Jaison R., Ernst R. Berndt, and Cory W. Monroe. 2004. "Hedonic Price Indexes for Personal Computer Operating Systems and Productivity Suites." NBER Working Paper No. 10427. April.

Aizcorbe, Ana, Kenneth Flamm, and Anjum Khurshid. 2002. "The Role of Semiconductor Inputs in IT Hardware Price Decline: Computers vs. Communications." Federal Reserve Board Finance and Economics Series Discussion Paper 2002-37. August.

Archibald, Robert B., and William S. Reece. 1979. "Partial Subindexes of Input Prices: The Case of Computer Services." *Southern Economic Journal* 46 (October):528-540.

Baily, M. N. and R. Z. Lawrence. 2001. "Do We Have an E-conomy?" NBER Working Paper 8243. April 23.

Bapco. 2002. "SYSmark® 2002: An Overview of SYSmark 2002 Business Applications Performance Corporation." Available at <*http://www.bapco.com/SYSmark2002Methodology.pdf*>, accessed February 19, 2003.

Bard, Yonathan and Charles H. Sauer. 1981. "IBM Contributions to Computer Performance Modeling." *IBM Journal of Research and Development* 25:562-570.

Barzyk, Fred. 1999. "Updating the Hedonic Equations for the Price of Computers." Working Paper of Statistics Canada, Prices Division, November 2.

Bell, C. Gordon. 1986. "RISC: Back to the Future?" *Datamation* 32 (June):96-108.

Benkard, C. Lanier. 2001. *A Dynamic Analysis of the Market for Wide Bodied Commercial Aircraft.* Stanford: Graduate School of Business, Stanford University. June.

Berndt, Ernst R. and Zvi Griliches. 1993. "Price Indexes for Microcomputers: An Exploratory Study." In Murray F. Foss, Marilyn Manser, and Allan H. Young, eds. *Price Measurements and Their Uses*. Studies in Income and Wealth 57:63-93. Chicago: University of Chicago Press for the National Bureau of Economic Research.

Berndt, Ernst R. and Neal J. Rappaport. 2001. "Price and Quality of Desktop and Mobile Personal Computers: A Quarter-Century Historical Overview." *American Economic Review* 91(2):268-273.

Berndt, Ernst R. and Neal J. Rappaport. 2002. "Hedonics for Personal Computers: A Reexamination of Selected Econometric Issues." Unpublished Paper.

Berndt, Ernst R., Zvi Griliches, and Neal Rappaport. 1995. "Econometric Estimates of Prices in Indexes for Personal Computers in the 1990s." *Journal of Econometrics* 68(1995):243-268.

Bloch, Erich and Dom Galage. 1978. "Component Progress: Its Effect on High-Speed Computer Architecture and Machine Organization." *Computer* 11 (April):64-75.

Boehm, Barry, Chris Abts, A. Brown, Sunita Chulani, Bradford Clark, and Ellis Horowitz. 2005. *Software Cost Estimation with Cocomo II*. Pearson Education.

Bonvillian, William B. 2004. "Meeting the New Challenge to U.S. Economic Competitiveness." *Issues in Science and Technology* XXI(1):75-82.

Bourot, Laurent. 1997. "Indice de Prix des Micro-ordinateurs et des Imprimantes: Bilan d'une rénovation." Working Paper of the Institut National De La Statistique Et Des Etudes Economiques (INSEE). Paris, France. March 12.

Brainard, Lael and Robert E. Litan. 2004. " 'Off-shoring' Service Jobs: Bane or Boon and What to Do?" Brookings Institution Policy Brief 132. April.

Brinkman, W. F. 1986. *Physics Through the* 1990s. Washington, D.C.: National Academy Press.

Bromley, D. Alan. 1972. *Physics in Perspective*, Washington, D.C.: National Academy Press.

Brooks, Frederick. 1975. *The Mythical Man Month: Essays on Software Engineering*. New York: Addison-Wesley Publishing Company.

Browning, L. D. and J. C. Shetler. 2000. *SEMATECH, Saving the U.S. Semiconductor Industry*. College Station: Texas A&M Press.

Brynjolfsson, Erik and Lorin M. Hitt. 2003. "Computing Productivity: Firm-Level Evidence." *Review of Economics and Statistics* 85(4):793-808.

Bureau of Economic Analysis. 2001. "A Guide to the NIPAs." In *National Income and Product Accounts of the United States, 1929-97*. Washington, D.C.: Government Printing Office. Also available at <*http://www.bea.doc.gov/bea/an/nipaguid.pdf*>.

Butler Group. 2001. "Is Clock Speed the Best Gauge for Processor Performance?" *Server World Magazine* September 2001. Available at <*http://www.serverworldmagazine.com/opinionw/2001/09/06_clockspeed.shtml*>, accessed February 7, 2003.

Cahners In-Stat Group. 1999. "Is China's Semiconductor Industry Market Worth the Risk for Multinationals? Definitely!" March 29.

Cale, E. G., L. L. Gremillion, and J. L. McKenney. 1979. "Price/Performance Patterns of U.S. Computer Systems." *Communications of the Association for Computing Machinery (ACM)* 22 (April):225-233.

Cartwright, David W. 1986. "Improved Deflation of Purchases of Computers." *Survey of Current Business* 66(3):7-9. March.

Cartwright, David W., Gerald F. Donahoe, and Robert P. Parker. 1985. "Improved Deflation of Computer in the Gross National Product of the United States." Bureau of Economic Analysis Working Paper 4. Washington, D.C.: U.S. Department of Commerce. December.

Cholewa, Rainier. 1996. "16M DRAM Manufacturing Cooperation IBM/SIEMENS in Corbeil Essonnes in France," *Proceedings of the 1996 IEEE/SEMI Advanced Semiconductor Manufacturing Conference*.

Chow, Gregory C. 1967. "Technological Change and the Demand for Computers." *American Economic Review* 57 (December):1117-1130.

Christiansen, Clayton. 1997. *The Innovator's Dilemma: When New Technologies Cause Great Firms to Fail.* Cambridge, MA: Harvard Business School Press.

Chwelos, Paul. 2003. "Approaches to Performance Measurement in Hedonic Analysis: Price Indexes for Laptop Computers in the 1990s." *Economics of Innovation and New Technology* 12(3):199-224.

Cohen, Stephen S. and John Zysman. 1988. *Manufacturing Matters: The Myth of the Post-Industrial Economy.* New York: Basic Books.

Cohen, Wesley M. and John Walsh. 2002. "Public Research, Patents and Implications for Industrial R&D in the Drug, Biotechnology, Semiconductor and Computer Industries." In National Research Council, *Capitalizing on New Needs and New Opportunities: Government-Industry Partnerships in Biotechnology and Information Technologies.* Washington, D.C.: National Academy Press.

Cole, Rosanne, Y. C. Chen, Joan A. Barquin-Stolleman, Ellen Dulberger, Nurhan Helvacian, and James H. Hodge. 1986. "Quality-Adjusted Price Indexes for Computer Processors and Selected Peripheral Equipment." *Survey of Current Business* 66(1):41-50. January.

Colecchia, Alessandra and Schreyer, Paul. 2002. "ICT Investment and Economic Growth in the 1990s: Is the United States a Unique Case? A Comparative Study of Nine OECD Countries." *Review of Economic Dynamics* 5(2):408-442.

Computerworld. 1999. "Generally Accepted Accounting Principles." November 29.

Computerworld. 2003. "Software Failure Cited in August Blackout Investigation." November 20.

Cunningham, Carl, Denis Fandel, Paul Landler, and Robert Wright. 2000. *Silicon Productivity Trends.* International SEMATECH Technology Transfer #00013875A-ENG. February 29.

Dalén, Jorgen. 1989. "Using Hedonic Regression for Computer Equipment in the Producer Price Index." R&D Report, Statistics Sweden, Research-Methods-Development.

David, Paul A. 2000. *Understanding the Digital Economy.* Cambridge, MA: MIT Press.

Dulberger, Ellen R. 1989. "The Application of a Hedonic Model to a Quality Adjusted Price Index for Computer Processors." In Dale W. Jorgenson and Ralph Landau, eds. *Technology and Capital Formation.* Cambridge, MA: MIT Press.

Dulberger, Ellen. 1993. "Sources of Price Decline in Computer Processors: Selected Electronic Components." In Murray Foss, Marilyn Manser, and Allan Young, eds. *Price Measurements and Their Uses.* Chicago: University of Chicago Press for the National Bureau of Economic Research.

The Economist. 2000. "A Thinker's Guide." March 30.

The Economist. 2001. "The Great Chip Glut." August 11.

The Economist. 2003. "The New 'New Economy.' " September 11.

The Economist. 2005. "Moore Law at 40." March 26.

Ein-Dor, Phillip. 1985. "Grosh's Law Re-visited: CPU Power and the Cost of Computation." *Communications of the Association for Computing Machinery (ACM)* 28(February):142-151.

Ericson, R. and A. Pakes. 1995. "Markov-Perfect Industry Dynamics: A Framework for Empirical Work." *Review of Economic Studies* 62:53-82.

Evans, Richard. 2002. "INSEE's Adoption of Market Intelligence Data for its Hedonic Computer Manufacturing Price Index." Presented at the Symposium on Hedonics at Statistics Netherlands, October 25.

Executive Office of the President. 2003. *The National Strategy to Secure Cyberspace: Cyberspace Threats and Vulnerabilities.* February.

Fershtman, C. and A. Pakes. 2000. "A Dynamic Game with Collusion and Price Wars." *RAND Journal of Economics* 31(2):207-236.

Fisher, Franklin M., John J. McGowan, and Joen E. Greenwood. 1983. *Folded, Spindled, and Multiplied: Economic Analysis and U.S. v. IBM.* Cambridge, MA: MIT Press.

Flamm, Kenneth. 1987. *Targeting the Computer.* Washington, D.C.: The Brookings Institution.

Flamm, Kenneth. 1988. *Creating the Computer.* Washington, D.C.: The Brookings Institution.

Flamm, Kenneth. 1989. "Technological Advance and Costs: Computers vs. Communications." In Robert C. Crandall and Kenneth Flamm, eds. *Changing the Rules: Technological Change, International Competition, and Regulation in Communications*. Washington, D.C.: The Brookings Institution.

Flamm, Kenneth. 1993. "Measurement of DRAM Prices: Technology and Market Structure." In Murray F. Foss, Marilyn E. Manser, and Allan H. Young, eds. *Price Measurements and Their Uses*. Chicago: University of Chicago Press.

Flamm, Kenneth. 1996. "Japan's New Semiconductor Technology Programs." *Asia Technology Information Program Report No. ATIP 96.091*. Tokyo. November.

Flamm, Kenneth. 1996. *Mismanaged Trade? Strategic Policy and the Semiconductor Industry*. Washington, D.C.: The Brookings Institution.

Flamm, Kenneth. 1997. *More For Less: The Economic Impact of Semiconductors*. San Jose, CA: Semiconductor Industry Association. December.

Fransman, M. 1992. *The Market and Beyond: Cooperation and Competition in Information Technology Development in the Japanese System*. Cambridge, UK: Cambridge University Press.

Gandal, Neil. 1994. "Hedonic Price Indexes for Spreadsheets and an Empirical Test for Network Externalities." *RAND Journal of Economics* 25.

Gandal, Neil and Chaim Fershtman. 2004. "The Determinants of Output per Contributor in Open Source Projects: An Empirical Examination." CEPR Working Paper 2650. Available at *<http://spirit.tau.ac.il/public/gandal/Research.htmworkingpapers/papers2/9900/00-059.pdf>*.

Gordon, Robert J. 1989. "The Postwar Evolution of Computer Prices." In Dale W. Jorgenson and Ralph Landau, eds. *Technology and Capital Formation*. Cambridge, MA: MIT Press.

Gosling, James, Bill Joy, and Guy Steele. 1996. *The Java (TM) Language Specification*. New York: Addison-Wesley.

Gowrisankaran, G. 1998. "Issues and Prospects for Payment System Deregulation." Working Paper, University of Minnesota.

Greenspan, Alan. 2000. Remarks before the White House Conference on the New Economy. Washington D.C., April 5. Available at *<http://www.federalreserve.gov/BOARDDOCS/SPEECHES/2000/20000405.HTM>*.

Griffith, P. 1993. "Science and the Public Interest." *The Bridge*. Washington, D.C.: National Academy of Engineering. Fall. p. 16.

Grindley, P., D. C. Mowery, and B. Silverman. 1994. "SEMATECH and Collaborative Research: Lessons in the Design of a High-Technology Consortia." *Journal of Policy Analysis and Management* 13.

Grossman, Gene and Elhannan Helpman. 1993. *Innovation and Growth in the Global Economy*. Cambridge, MA: MIT Press.

Hagel, J. and A. G. Armstrong. 1997. *Net Gain*. Cambridge, MA: Harvard Business School Press.

Halloran, T. J. and William Scherlis. 2002. "High Quality and Open Source Software Practices." Position Paper. 24th International Conference on Software Engineering.

Halstead, Maurice H. 1977. *Elements of Software Science*. New York: Elsevier North Holland.

Handler, Philip. 1970. *Biology and the Future of Man*. London: Oxford University Press.

Harhoff, Dietmar and Dietmar Moch. 1997. "Price Indexes for PC Database Software and the Value of Code Compatibility." *Research Policy* 24(4-5):509-520.

Holdway, Michael. 2001. "Quality-Adjusting Computer Prices in the Producer Price Index: An Overview." Washington, D.C.: Bureau of Labor Statistics. October 16.

Horrigan, John Brendan. 1996. "Cooperation Among Competitors in Research Consortia." Unpublished doctoral dissertation, University of Texas at Austin. December.

Howell, Thomas. 2003. "Competing Programs: Government Support for Microelectronics." In National Research Council, *Securing the Future: Regional and National Programs to Support the Semiconductor Industry*. Charles W. Wessner, ed. Washington, D.C.: National Academies Press.

Ishida, Haruhisa. 1972. "On the Origin of the Gibson Mix." *Journal of the Information Processing Society of Japan* 13(May):333-334 (in Japanese).

Jorgenson, Dale W. 2001. "Information Technology and the U.S. Economy." *American Economic Review* 91(1). March.

Jorgenson, Dale W. 2002. "The Promise of Growth in the Information Age." The Conference Board, Annual Essay.

Jorgenson, Dale W. and Kevin J. Stiroh. 1999. "Productivity Growth: Current Recovery and Longer-term Trends." *American Economic Review* 89(2).

Jorgenson, Dale W. and Kevin J. Stiroh. 2002. "Raising the Speed Limit: U.S. Economic Growth in the Information Age." In National Research Council. *Measuring and Sustaining the New Economy.* Dale W. Jorgenson and Charles W. Wessner, eds. Washington, D.C.: National Academy Press.

Jorgenson, Dale W., Kevin J. Stiroh, Robert J. Gordon, and Daniel E. Sichel. 2000. "Raising the Speed Limit: U.S. Economic Growth in the Information Age." *Brookings Papers on Economic Activity* 2000(1):125-235.

Jorgenson, Dale W., Mun S. Ho, and Kevin J. Stiroh. 2002. "Information Technology, Education, and the Sources of Economic Growth Across U.S. Industries." Presented at the Brookings Workshop "Services Industry Productivity: New Estimates and New Problems," March 14. Available at <*http://www.brook.edu/dybdocroot/es/research/projects/productivity/ workshops/20020517.htm*>.

Jorgenson, Dale W., Mun S. Ho, and Kevin J. Stiroh. 2004. "Will the U.S. Productivity Resurgence Continue?" *FRBNY Current Issues in Economics and Finance* 10(1).

Kelejian, Harry H. and Robert V. Nicoletti. c. 1971. "The Rental Price of Computers: An Attribute Approach." Unpublished Paper, New York University (no date).

Kessler, Michelle. 2002. "Computer Majors Down Amid Tech Bust." *USA Today* October 8.

Knight, Kenneth E. 1966. "Changes in Computer Performance: A Historical View." *Datamation* (September):40-54.

Knight, Kenneth E. 1970. "Application of Technological Forecasting to the Computer Industry." In James R. Bright and Milton E.F. Schieman. *A Guide to Practical Technological Forecasting.* Englewood Cliffs, NJ: Prentice-Hall.

Knight, Kenneth E. 1985. "A Functional and Structural Measure of Technology." *Technological Forecasting and Technical Change* 27(May):107-127.

Koskimäki, Timo and Yrjö Vartia. 2001. "Beyond Matched Pairs and Griliches-type Hedonic Methods for Controlling Quality Changes in CPI Sub-indices." Presented at the Sixth Meeting of the International Working Group on Price Indices, sponsored by the Australian Bureau of Statistics, April.

Lam, Monica. 2003. Private communication. November 7.

Lerner, Josh and Jean Tirole. 2000. "The Simple Economics of Open Source." Harvard Business School. February 25. Available at <*http://www.hbs.edu/research/facpubs/*>.

Levine, Jordan. 2002. "U.S. Producer Price Index for Pre-Packaged Software." Presented at the 17th Voorburg Group Meeting. Nantes, France. September.

Levy, David and Steve Welzer. 1985. "An Unintended Consequence of Antitrust Policy: The Effect of the IBM Suit on Pricing Policy." Unpublished Paper, Rutgers University Department of Economics. December.

Lias, Edward. 1980. "Tacking the Elusive KOPS." *Datamation* (November):99-118.

Lim, Poh Ping and Richard McKenzie. 2002. "Hedonic Price Analysis for Personal Computers in Australia: An Alternative Approach to Quality Adjustments in the Australian Price Indexes."

Lipson, Howard F. 2002. "Tracking and Tracing Cyber-Attacks: Technical Challenges and Global Policy Issues." CERT Coordination Center. CMU/SEI2002-SR-009. November.

Macher, Jeffrey T., David C. Mowery, and David A. Hodges. 1999. "Semiconductors." In National Research Council. *U.S. Industry in 2000: Studies in Competitive Performance.* David C. Mowery, ed. Washington, D.C.: National Academy Press.

Maney, Kevin. 2003. "Music Industry Doesn't Know What Else To Do As It Lashes Out at File-sharing." *USA Today* September 9.

Mann, Catherine. 2003. "Globalization of IT Services and White Collar Jobs: The Next Wave of Productivity Growth." *International Economics Policy Briefs* PB03-11. December.

Martin, Brookes and Zaki Wahhaj. 2000. "The Shocking Economic Impact of B2B." *Global Economic Paper* 37. Goldman Sachs. February 3, 2000.

McKinsey Global Institute. 2001. *U.S. Productivity Growth 1995-2000, Understanding the Contribution of Information Technology Relative to Other Factors.* Washington, D.C.: McKinsey & Co. October.

McKinsey Global Institute. 2003. *Offshoring: Is it a win-win game?* San Francisco: McKinsey Global Institute.

Meller, Paul. 2003. "European Parliament Votes to Limit Software Patents." *New York Times* September 25.

Michaels, Robert. 1979. "Hedonic Prices and the Structure of the Digital Computer Industry." *The Journal of Industrial Economics* 27(March):263-275.

Moch, Dietmar. 2001. "Price Indices for Information and Communication Technology Industries: An Application to the German PC Market." Center for European Economic Research (ZEW) Discussion Paper No. 01-20. Mannheim, Germany. August.

Mockus, A., R. Fielding, and J. D. Herbsleb. 2002. "Two Case Studies of Open Source Software Development: Apache and Mozilla." *ACM Transactions on Software Engineering and Methodology* 11(3):309-346.

Moore, Gordon E. 1965. "Cramming More Components onto Integrated Circuits." *Electronics* 38(8). April.

Moore, Gordon E. 1975. "Progress in Digital Integrated Circuits." *Proceedings of the 1975 International Electron Devices Meeting.* pp. 11-13.

Moore, Gordon E. 1997. "The Continuing Silicon Technology Evolution Inside the PC Platform." *Intel Developer Update* Issue 2. October 15.

Mossberg, Walter. 2004. "Mossberg's Mailbox." *Wall Street Journal* June 24.

Moylan, Carol, 2001. "Estimation of Software in the U.S. National Income and Product Accounts: New Developments." OECD Paper. September 2001. Available at <*http://webnet1.oecd.org/doc/M00017000/M00017821.doc*>.

National Academy of Sciences, National Academy of Engineering, Institute of Medicine. 1993. *Science, Technology and the Federal Government. National Goals for a New Era.* Washington, D.C.: National Academy Press.

National Advisory Committee on Semiconductors. 1989. *A Strategic Industry at Risk.* Washington, D.C.: National Advisory Committee on Semiconductors.

National Advisory Committee on Semiconductors. 1990. *Capital Investment in Semiconductors: The Lifeblood of the U.S. Semiconductor Industry.* Washington, D.C.: National Advisory Committee on Semiconductors. September.

National Advisory Committee on Semiconductors. 1990. *Preserving the Vital Base: America's Semiconductor Materials and Equipment Industry.* Washington, D.C.: National Advisory Committee on Semiconductors. July.

National Advisory Committee on Semiconductors. 1991. *MICROTECH 2000 Workshop Report.* Washington, D.C.: National Advisory Committee on Semiconductors.

National Advisory Committee on Semiconductors. 1991. *Toward a National Semiconductor Strategy.* Vols. 1 and 2. Washington, D.C.: National Advisory Committee on Semiconductors. February.

National Advisory Committee on Semiconductors. 1992. *Competing in Semiconductors.* Washington, D.C.: National Advisory Committee on Semiconductors. August.

National Advisory Committee on Semiconductors. 1992. *A National Strategy for Semiconductors: An Agenda for the President, the Congress, and the Industry.* Washington, D.C.: National Advisory Committee on Semiconductors. February.

National Research Council. 1996. *Conflict and Cooperation in National Competition for High-Technology Industry*. Washington, D.C.: National Academy Press.

National Research Council. 1999. *The Advanced Technology Program: Challenges and Opportunities.* Charles W. Wessner, ed. Washington, D.C.: National Academy Press.

National Research Council. 1999. *Industry-Laboratory Partnerships: A Review of the Sandia Science and Technology Park Initiative.* Charles W. Wessner, ed. Washington, D.C.: National Academy Press.

National Research Council. 1999. *New Vistas in Transatlantic Science and Technology Cooperation.* Charles W. Wessner, ed. Washington, D.C.: National Academy Press.

National Research Council. 1999. *The Small Business Innovation Research Program: Challenges and Opportunities.* Charles W. Wessner, ed. Washington, D.C.: National Academy Press.

National Research Council. 1999. *U.S. Industry in 2000: Studies in Competitive Performance.* David C. Mowery, ed. Washington, D.C.: National Academy Press.

National Research Council. 2001. *The Advanced Technology Program: Assessing Outcomes.* Charles W. Wessner, ed. Washington, D.C.: National Academy Press.

National Research Council. 2001. *Capitalizing on New Needs and New Opportunities: Government-Industry Partnerships in Biotechnology and Information Technologies.* Charles W. Wessner, ed. Washington, D.C.: National Academy Press.

National Research Council. 2001. *A Review of the New Initiatives at the NASA Ames Research Center.* Charles W. Wessner, ed. Washington, D.C.: National Academy Press.

National Research Council. 2001. *Trends in Federal Support of Research and Graduate Education.* Stephen A. Merrill, ed. Washington, D.C.: National Academy Press.

National Research Council. 2002. *Government-Industry Partnerships for the Development of New Technologies: Summary Report.* Charles W. Wessner, ed. Washington, D.C.: National Academy Press.

National Research Council. 2002. *Measuring and Sustaining the New Economy.* Dale W. Jorgenson and Charles W. Wessner, eds. Washington, D.C.: National Academy Press.

National Research Council. 2003. *Innovation in Information Technology.* Washington, D.C.: National Academies Press.

National Research Council. 2003. *Securing the Future; Regional and National Programs to Support the Semiconductor Industry.* Charles W. Wessner, ed. Washington, D.C.: The National Academies Press.

National Research Council. 2004. *Productivity and Cyclicality in Semiconductors.* Dale W. Jorgenson and Charles W. Wessner, eds. Washington, D.C.: The National Academies Press.

National Research Council. 2005. *Deconstructing the Computer.* Dale W. Jorgenson and Charles W. Wessner, eds. Washington, D.C.: The National Academies Press.

National Research Council. Forthcoming. *The Telecommunications Challenge: Changing Technologies and Evolving Policies.* Dale W. Jorgenson and Charles W. Wessner, eds. Washington, D.C.: The National Academies Press.

Nelson, Richard, ed. 1993. *National Innovation Systems.* New York: Oxford University Press.

Nelson, R. A., T. L. Tanguay, and C. C. Patterson. 1994. "A Quality-adjusted Price Index for Personal Computers." *Journal of Business and Economics Statistics* 12(1):23-31.

Nordhaus, William D. 2002. "The Progress of Computing." Yale University. March 4.

Nikkei Microdevices. 2001. "From Stagnation to Growth, The Push to Strengthen Design." January.

Nikkei Microdevices. 2001. "Three Major European LSI Makers Show Stable Growth Through Large Investments." January.

Office of Senator Lieberman. 2003. White Paper: *National Security Aspects of the Global Migration of the U.S. Semiconductor Industry.* June.

Office of Senator Lieberman. 2004. White Paper: *Offshore Outsourcing and America's Competitive Edge: Losing Out in the High Technology R&D and Services Sectors.* May 11.

Office of Senator Lieberman. 2004. *Data Dearth in Offshore Outsourcing: Policymaking Requires Facts*. December.

O'Gara, Maureen. 2004. "Huge MyDoom Zombie Army Wipes Out SCO." *Linux World Magazine* February 1.

Ohta, Makoto and Zvi Griliches. 1976. "Automobile Prices Revisited: Extensions of the Hedonic Hypothesis." In Nestor E. Terleckyj, ed. *Household Production and Consumption*. Conference on Research in Income and Wealth. *Studies in Income and Wealth* 40:325-90. New York: National Bureau of Economic Research.

Okamoto, Masato and Tomohiko Sato. 2001. "Comparison of Hedonic Method and Matched Models Method Using Scanner Data: The Case of PCs, TVs and Digital Cameras." Presented at Sixth Meeting of the International Working Group on Price Indices, sponsored by the Australian Bureau of Statistics, April.

Oliner, Stephen and Daniel Sichel. 2000. "The Resurgence of Growth in the Late 1990s: Is Information Technology the Story?" *Journal of Economic Perspectives* 14(4). Fall.

Organisation for Economic Co-operation and Development. 2000. *Is There a New Economy? A First Report on the OECD Growth Project.* Paris: Organisation for Economic Co-operation and Development. June.

Organisation for Economic Co-operation and Development. 2002. *Information Technology Outlook 2002—The Software Sector.* Paris: Organisation for Economic Co-operation and Development.

Organisation for Economic Co-operation and Development. 2002. *Strengthening the Knowledge-based Economy*. Paris: Organisation for Economic Co-operation and Development.

Organisation for Economic Co-operation and Development. 2003. *ICT and Economic Growth: Evidence from OECD Countries, Industries and Firms*. Paris: Organisation for Economic Co-operation and Development.

Organisation for Economic Co-operation and Development. 2003. Statistics Working Paper 2003/1: Report of the OECD Task Force on Software Measurement in the National Accounts. Paris: Organisation for Economic Co-operation and Development.

Pake, G. E. 1966. *Physics Survey and Outlook.* Washington, D.C.: National Academy Press.

Pakes, Ariel. 2001. "A Reconsideration of Hedonic Price Indices with an Application to PCs." Harvard University, November.

Parker, Robert P. and Bruce Grimm. 2000. "Recognition of Business and Government Expenditures for Software as Investment: Methodology and Quantitative Impacts, 1959-98." Paper presented to Bureau of Economic Analysis's Advisory Committee, May 5, 2000. Available at <*http://www.bea.doc.gov/bea/papers/software.pdf*>.

Patrick, James M. 1969. "Computer Cost/Effectiveness." Unpublished Paper summarized in Sharpe 1969. p. 352.

PC World. 2003. "20 Years of Hardware." March.

Phister, Montgomery. 1979. *Data Processing Technology and Economics.* Second Edition. Bedford, MA: Santa Monica Company Publishing and Digital Press.

Porter, Michael. 2004. "Building the Microeconomic Foundations of Prosperity: Findings from the Business Competitiveness Index." In X Sala-i-Martin, ed. *The Global Competitiveness Report 2003-2004*. New York: Oxford University Press.

Poulsen, Kevin. 2004. "Software Bug Contributed to Blackout." *Security Focus* February 11.

Prince, Betty. 1991. *Semiconductor Memories: A Handbook of Design, Manufacture and Application.* Second Edition. Chichester, UK: John Wiley and Sons.

Pritchard, Stephen. 2003. "Munich Makes the Move." *Financial Times* October 15.

Prud'homme, Marc and Kam Yu. 2002. "A Price Index for Computer Software Using Scanner Data." Unpublished Working Paper. Prices Division, Statistics Canada.

Rao, H. Raghaw and Brian D. Lynch. 1993. "Hedonic Price Analysis of Workstation Attributes." *Communications of the Association for Computing Machinery (ACM)* 36(12):94-103.

Ratchford, Brian T., and Gary T. Ford. 1976. "A Study of Prices and Market Shares in the Computer Mainframe Industry." *The Journal of Business* 49:194-218.

Ratchford, Brian T., and Gary T. Ford. 1979. "Reply." *The Journal of Business* 52:125-134.

Robertson, Jack. 1998. "Die Shrinks Now Causing Logic Chip Glut." *Semiconductor Business News* October 15.

Rosch, Winn L. 1994. *The Winn L. Rosch Hardware Bible*. Indianapolis: Sams Publishing.

Samuelson, Paul. 2004. "Where Ricardo and Mill Rebut and Confirm Arguments of Mainstream Economists Supporting Globalization." *Journal of Economic Perspectives* 18(3). Summer.

Schaller, Robert R. 1999. "Technology Roadmaps: Implications for Innovation, Strategy, and Policy." Ph.D. Dissertation Proposal, Institute for Public Policy, George Mason University.

Schaller, Robert R. 2002. "Moore's Law: Past, Present, and Future." Available at <*http://www.njtu.edu.cn/depart/xydzxx/ec/spectrum/moore/mlaw.html*>. Accessed July 2002.

Schlender, Brent. 1999. "The Edison of the Internet." *Fortune* February 15.

Semiconductor Industry Association. 1993. *Semiconductor Technology Workshop Conclusions*. San Jose, CA: Semiconductor Industry Association.

Semiconductor Industry Association. 1993. *Semiconductor Technology Workshop Working Group Reports*. San Jose, CA: Semiconductor Industry Association.

Semiconductor Industry Association. 1994. *The National Technology Roadmap for Semiconductors, 1994*. San Jose, CA: Semiconductor Industry Association.

Semiconductor Industry Association. 1997. *The National Technology Roadmap for Semiconductors, Technology Needs*. 1997 Edition. San Jose, CA: Semiconductor Industry Association.

Serlin, Omri. 1986. "MIPS, Dhrystones, and Other Tables." *Datamation* 32(June 1):112-118.

Sharpe, William F. 1969. *The Economics of the Computer*. New York and London: Columbia University Press.

Shuster, Loren. 2002. "Global Gaming Industry Now a Whopping $35 Billion Market." *Compiler* July 2002.

Sigurdson, J. *Industry and State Partnership in Japan: The Very Large Scale Integrated Circuits (VLSI) Project*. Lund, Sweden: Research Policy Institute.

Solow, Robert M. 1987. "We'd Better Watch Out." *New York Times Book Review* July 12.

Solow, Robert M., Michael Dertouzos, and Richard Lester. 1989. *Made in America*. Cambridge, MA: MIT Press.

Spencer, W. J. and P. Grindley 1993. "SEMATECH After Five Years: High Technology Consortia and U.S. Competitiveness." *California Management Review* 35.

Stapper, Charles H. and Raymond J. Rosner. 1995. "Integrated Circuit Yield Management and Yield Analysis: Development and Implementation." *IEEE Transactions on Semiconductor Manufacturing* 8(2). May.

Statistics Finland. 2000. "Measuring the Price Development of Personal Computers in the Consumer Price Index." Paper for the Meeting of the International Hedonic Price Indexes Project. Paris, France. September 27.

Stoneman, Paul. 1976. *Technological Diffusion and the Computer Revolution: The U.K. Experience*. Cambridge, UK: Cambridge University Press.

Stoneman, Paul. 1978. "Merger and Technological Progressiveness: The Case of the British Computer Industry." *Applied Economics* 10:125-140. Reprinted as Chapter 9 in Keith Cowling, Paul Stoneman, John Cubbin, John Cable, Graham Hall, Simon Domberger, and Patricia Dutton. 1980. *Mergers and Economic Performance*. Cambridge, UK: Cambridge University Press.

Triplett, Jack E. 1985. "Measuring Technological Change with Characteristics-Space Techniques." *Technological Forecasting and Social Change* 27:283-307.

Triplett, Jack E. 1989. "Price and Technological Change in a Capital Good: A Survey of Research on Computers." In Dale W. Jorgenson and Ralph Landau, eds. *Technology and Capital Formation*. Cambridge, MA: MIT Press.

Triplett, Jack E. 1996. "High-Tech Productivity and Hedonic Price Indexes." In Organisation for Ecoomic Co-operation and Development. *Industry Productivity*. Paris: Organization for Economic Co-operation and Development.

Triplett, Jack E. and Barry Bosworth. 2002. "Baumol's Disease Has Been Cured: IT and Multifactor Productivity in U.S. Service Industries." Presented at the Brookings Workshop "Services Industry Productivity: New Estimates and New Problems," March 14. Available at *<http://www.brook.edu/dybdocroot/es/research/projects/productivity/workshops/20020517.htm>*.

van Mulligen, Peter Hein. 2002. "Alternative Price Indices for Computers in the Netherlands Using Scanner Data." Prepared for the 27th General Conference of the International Association for Research in Income and Wealth. Djurhamn, Sweden.

Varian, Hal R. and Carl Shapiro. 2003. "Linux Adoption in the Public Sector: An Economic Analysis." Department of Economics. University of California at Berkeley. December 1. Available at *<http://www.sims.berkeley.edu/~hal/Papers/ 2004/linux-adoption-in-the-public-sector.pdf>*.

VeriTest. 2003. "Business Winstone™ 2002 Basics." Available at *<http://www.veritest.com/benchmarks/bwinstone/wshome.asp>*, accessed February 19, 2003.

Wallace, William E. 1985. "Industrial Policies and the Computer Industry." The Futures Group Working Paper #007. Glastonbury, CT: The Futures Group. March.

The Washington Post. 2004. "Election Campaign Hit More Sour Notes." Page F-02. February 22.

Wolff, Alan Wm., Thomas R. Howell, Brent L. Bartlett, and R. Michael Gadbaw, eds. 1992. *Conflict Among Nations: Trade Policies in the 1990s*. San Francisco: Westview Press.

World Semiconductor Trade Statistics. 2000. *Annual Consumption Survey*.

Wyckoff, Andrew W. 1995. "The Impact of Computer Prices on International Comparisons of Labour Productivity." *Economics of Innovation and New Technology* 3:277-293.